7th Grade PSSA

Math Exercise Book

Review of Essential Skills and Concepts

With 2 PSSA Math Practice Tests

By

Elise Baniam & Michael Smith

7th Grade PSSA Math Exercise Book

Published in the United State of America By

The Math Notion

Email: info@mathnotion.com

Web: www.mathnotion.com

ISBN: 978-1-63620-131-3

About the Author

Elise Baniam has been a math instructor for over a decade now. She graduated in Mathematics. Since 2006, Elise has devoted his time to both teaching and developing exceptional math learning materials. As a Math instructor and test prep expert, Elise has worked with thousands of students. She has used the feedback of her students to develop a unique study program that can be used by students to drastically improve their math score fast and effectively.

- **ACT Math Workbook**
- **SAT Math Workbook**
- **ISEE/SSAT Math Workbook**
- **Common Core Math Workbook**
- **many Math Education Workbooks**
- **and some Mathematics books …**

As an experienced Math teacher, Mrs. Baniam employs a variety of formats to help students achieve their goals: she teaches students in large groups, and she provides training materials and textbooks through her website and through Amazon.

You can contact Elise via email at:

Elise@mathnotion.com

Get All the Math Prep You Need to Ace the 7ᵗʰ Grade PSSA Test!

Studying for a test is much easier when you know what will be on it, particularly when you can crack it down into apparent parts. You can then study each section independently.

7ᵗʰ Grade PSSA Math Exercise Book helps you achieve the next level of professional achievement. It contains over 2,500 practice problems covering every topic tested on the PSSA math grade 7, making it a critical resource for students to provide them with comprehensive practice.

Upgraded by our professional instructors, the problems are sensibly categorized into practice sets and reflect those found on the math PSSA grade 7 in content, form, and style. Students can build fundamental skills in math through targeted practice while easy-to-follow explanations help cement their understanding of the concepts assessed on the PSSA math 7ᵗʰ grade.

This user-friendly resource includes simple explanations:

- Hands-on experience with all PSSA 7ᵗʰ grade math questions.
- Focusing your study time on what is most important.
- Everything you need to know for a High Score.
- Complete review to help you master different concepts.
- These reviews go into detail to cover all math topics on the PSSA 7ᵗʰ grade math.
- Hundreds of realistic questions and drills, including new practice questions.
- **2 full-length practice tests** with detailed answer explanations

Effective exercises to help you avoid traps and pacing yourself beat the PSSA math grade 7. It is packed with everything you need to do your best on the test and move toward your graduation.

WWW.MATHNOTION.COM

… So Much More Online!

✓ FREE Math Lessons

✓ More Math Learning Books!

✓ Mathematics Worksheets

✓ Online Math Tutors

For a PDF Version of This Book

Please Visit www.mathnotion.com

Contents

Chapter 1 : Whole Numbers 11

Add and Subtract Integers 12

Multiplication and Division 13

Ordering Integers and Numbers 14

Order of Operations 15

Factoring 16

Great Common Factor (GCF) 17

Least Common Multiple (LCM) 18

Divisibility Rule 19

Answers of Worksheets 20

Chapter 2 : Fractions 23

Adding Fractions – Unlike Denominator 24

Subtracting Fractions – Unlike Denominator 25

Converting Mix Numbers 26

Converting improper Fractions 27

Addition Mix Numbers 28

Subtracting Mix Numbers 29

Simplify Fractions 30

Multiplying Fractions 31

Multiplying Mixed Number 32

Dividing Fractions 33

Dividing Mixed Number 34

Comparing Fractions 35

Answers of Worksheets 36

Chapter 3 : Decimal 41

Round Decimals 42

Decimals Addition 43

Decimals Subtraction 44

Decimals Multiplication 45

Decimal Division 46

Comparing Decimals 47

Convert Fraction to Decimal 48

Convert Decimal to Percent 49

Convert Fraction to Percent 50

Answers of Worksheets 51

Chapter 4 : Geometry **54**

Angles 55

Area and Perimeter of Trapezoid 56

Area and Perimeter of Parallelogram 57

Circumference and Area of Circle 58

Perimeter of Polygon 59

Volume of Cubes 60

Volume of Rectangle Prism 61

Volume of Cylinder 62

Volume of Pyramid and Cone 63

Surface Area Cubes 64

Surface Area Rectangle Prism 65

Surface Area Cylinder 66

Answers of Worksheets 67

Chapter 5 : Exponent and Radicals **69**

Positive Exponents 70

Negative Exponents 71

Add and subtract Exponents 72

Exponent multiplication 73

Exponent division 74

Scientific Notation 75

Square Roots 76

Simplify Square Roots 77

Answers of Worksheets 78

Chapter 6 : Ratio, Proportion and Percent **81**

Proportions 82

Reduce Ratio 83

Percent 84

Discount, Tax and Tip 85

Percent of Change 86

Simple Interest 87

Answers of Worksheets 88

Chapter 7 : Transformations **90**

Translations 91

Reflections.. 92

Rotations.. 94

Dilations 96

Coordinates of Vertices 97

Answers of Worksheets 98

Chapter 8 : Equations and Inequality **101**

Distributive and Simplifying Expressions 102

Factoring Expressions 103

Evaluate One Variable Expressions 104

Evaluate Two Variable Expressions 105

Graphing Linear Equation 106

One Step Equations... 107

Two Steps Equations 108

Multi Steps Equations 109

Graphing Linear Inequalities... 110

One Step Inequality 111

Two Steps Inequality.. 112

Multi Steps Inequality.. 113

Systems of Equations 114

Systems of Equations Word Problems 115

Finding Distance of Two Points 116

Answers of Worksheets 117

Chapter 9 : Linear Functions... **123**

Relation and Functions... 124

Slope form... 125

Slope and Y-Intercept 125

Slope and One Point 126

Slope of Two Points.. 127

Equation of Parallel and Perpendicular lines 128

Answers of Worksheets 129

Chapter 10 : Statistics and probability **131**

Mean, Median, Mode, and Range of the Given Data 132

Box and Whisker Plot 133

Histogram 134

Dot plots 135

Scatter Plots 136

Pie Graph 137

Probability 138

Answers of Worksheets 139

Chapter 11 : PSSA Mathematics Test Review **143**

Grade 7 PSSA Mathematics Formula Sheet 145

PSSA Practice Test 1 147

PSSA Practice Test 2 157

Chapter 12 : Answers and Explanations **167**

Answer Key 169

PSSA Practice Test 1 171

PSSA Practice Test 2 175

Chapter 1 :

Whole Numbers

Add and Subtract Integers

Find the sum or difference.

1) $(+142) + (-88) =$

2) $(+64) + (-32) =$

3) $288 - 185 =$

4) $(-214) + 157 =$

5) $(-72) + 425 =$

6) $182 + (-265) =$

7) $(-15) + 38 =$

8) $415 - 310 =$

9) $(-18) - (-77) =$

10) $(-88) + (-57) =$

11) $(-124) - 304 =$

12) $1,520 - (-157) =$

13) $24 + (-20) + (-45) + (-20) =$

14) $(-27) + (-24) + 52 + 12 =$

15) $(-5) - 7 + 38 - 21 =$

16) $8 + (-19) + (-29 - 21) =$

17) $(+145) + (+28) + (-157) =$

18) $(-42) + (-32) =$

19) $-14 - 18 - 9 - 31 =$

20) $9 + (-28) =$

21) $134 - 90 - 53 - (-42) =$

22) $(+37) - (-9) =$

23) $(+17) - (+21) - (-19) =$

24) $(+37) - (+9) - (-42) =$

Multiplication and Division

Calculate.

1) $240 \times 7 =$

2) $130 \times 20 =$

3) $(-5) \times 7 \times (-4) =$

4) $-7 \times (-6) \times (-6) =$

5) $11 \times (-11) =$

6) $80 \times (-4) =$

7) $2 \times (-6) \times 8 =$

8) $(-300) \times (-40) =$

9) $(-30) \times (-20) \times 2 =$

10) $115 \times 5 =$

11) $142 \times 50 =$

12) $364 \div 14 =$

13) $(-4,125) \div 5 =$

14) $(-28) \div (-7) =$

15) $288 \div (-18) =$

16) $3,500 \div 28 =$

17) $(-126) \div 3 =$

18) $4,128 \div 4 =$

19) $1,260 \div (-35) =$

20) $3,360 \div 4 =$

21) $(-58) \div 2 =$

22) $(-10,000) \div (-50) =$

23) $0 \div 670 =$

24) $(-1,020) \div 6 =$

25) $5,868 \div 652 =$

26) $(-2,520) \div 4 =$

27) $10,902 \div 3 =$

28) $(-60) \div (-2) =$

Ordering Integers and Numbers

Order each set of integers from least to greatest.

1) $16, -10, -5, -1, 2$ ___, ___, ___, ___, ___, ___

2) $-14, -8, 15, 4, 11$ ___, ___, ___, ___, ___, ___

3) $39, -28, -16, 37, -21$ ___, ___, ___, ___, ___, ___

4) $-15, -55, 35, -27, 48$ ___, ___, ___, ___, ___, ___

5) $47, -32, 42, -35, 28$ ___, ___, ___, ___, ___, ___

6) $85, 38, -59, 95, -24$ ___, ___, ___, ___, ___, ___

Order each set of integers from greatest to least.

7) $16, 29, -21, -25, -4$ ___, ___, ___, ___, ___, ___

8) $12, 36, -54, -26, 71$ ___, ___, ___, ___, ___, ___

9) $55, -46, -19, 37, -17$ ___, ___, ___, ___, ___, ___

10) $37, 95, -46, -22, 87$ ___, ___, ___, ___, ___, ___

11) $-9, 79, -65, -78, 84$ ___, ___, ___, ___, ___, ___

12) $-80, -45, -60, 19, 39$ ___, ___, ___, ___, ___, ___

Order of Operations

Evaluate each expression.

1) $8 + (5 \times 2) =$

2) $25 - (6 \times 3) =$

3) $(15 \times 4) + 15 =$

4) $(21 - 6) - (5 \times 4) =$

5) $32 + (30 \div 5) =$

6) $(36 \times 8) \div 12 =$

7) $(63 \div 7) \times (-3) =$

8) $(7 \times 8) + (34 - 18) =$

9) $80 + (3 \times 3) + 5 =$

10) $(20 \times 8) \div (4 + 4) =$

11) $(-10) + (12 \times 5) + 14 =$

12) $(5 \times 7) - (35 \div 5) =$

13) $(7 \times 30 \div 10) - (17 + 13) =$

14) $(14 + 6 - 16) \times 8 - 12 =$

15) $(36 - 18 + 30) \times (96 \div 8) =$

16) $24 + \left(14 - (36 \div 6)\right) =$

17) $(7 + 10 - 4 - 9) + (24 \div 3) =$

18) $(90 - 15) + (16 - 18 + 8) =$

19) $(30 \times 3) + (16 \times 4) - 80 =$

20) $11 + 16 - (21 \times 3) + 25 =$

Factoring

Factor, write prime if prime.

1) 15

2) 41

3) 32

4) 85

5) 62

6) 65

7) 38

8) 10

9) 54

10) 121

11) 45

12) 90

13) 24

14) 33

15) 92

16) 57

17) 86

18) 40

19) 105

20) 80

21) 95

22) 81

23) 126

24) 110

25) 34

26) 98

27) 72

28) 104

Great Common Factor (GCF)

Find the GCF of the numbers.

1) 2, 18

2) 36, 23

3) 45, 35

4) 20, 32

5) 36, 64

6) 32, 42

7) 60, 25

8) 90, 35

9) 72, 9

10) 45, 54

11) 66, 54

12) 35, 70

13) 140, 40

14) 32, 82

15) 48, 96

16) 30, 85

17) 16, 24

18) 80, 100, 40

19) 81, 112

20) 56, 88

21) 20, 10, 50

22) 2, 9, 12

23) 30, 90, 120

24) 51, 33

Least Common Multiple (LCM)

Find the LCM of each.

1) 16, 20

2) 64, 32

3) 10, 20, 30

4) 28, 42

5) 10, 2, 15

6) 25, 5

7) 24, 120, 48

8) 15, 18

9) 13, 26, 54

10) 28, 35

11) 27, 54

12) 220, 44

13) 60, 30, 120

14) 36, 126

15) 40, 8, 5

16) 27, 6

17) 38, 19

18) 22, 44

19) 25, 60

20) 16, 48

21) 34, 20

22) 24, 28

23) 70, 140

24) 63, 18

Divisibility Rule

Apply the divisibility rules to find the factors of each number.

1) 16   2, 3, 4, 5, 6, 9, 10   13) 38   2, 3, 4, 5, 6, 9, 10

2) 121   2, 3, 4, 5, 6, 9, 10   14) 185   2, 3, 4, 5, 6, 9, 10

3) 252   2, 3, 4, 5, 6, 9, 10   15) 905   2, 3, 4, 5, 6, 9, 10

4) 74   2, 3, 4, 5, 6, 9, 10   16) 157   2, 3, 4, 5, 6, 9, 10

5) 241   2, 3, 4, 5, 6, 9, 10   17) 540   2, 3, 4, 5, 6, 9, 10

6) 155   2, 3, 4, 5, 6, 9, 10   18) 340   2, 3, 4, 5, 6, 9, 10

7) 65   2, 3, 4, 5, 6, 9, 10   19) 480   2, 3, 4, 5, 6, 9, 10

8) 320   2, 3, 4, 5, 6, 9, 10   20) 2,750   2, 3, 4, 5, 6, 9, 10

9) 1,134   2, 3, 4, 5, 6, 9, 10   21) 330   2, 3, 4, 5, 6, 9, 10

10) 68   2, 3, 4, 5, 6, 9, 10   22) 346   2, 3, 4, 5, 6, 9, 10

11) 754   2, 3, 4, 5, 6, 9, 10   23) 108   2, 3, 4, 5, 6, 9, 10

12) 128   2, 3, 4, 5, 6, 9, 10   24) 656   2, 3, 4, 5, 6, 9, 10

Answers of Worksheets

Add and Subtract Integers

1) 54	9) 59	17) 16
2) 32	10) −145	18) −74
3) 103	11) −428	19) −72
4) −57	12) 1,677	20) −19
5) 353	13) −61	21) 33
6) −83	14) 13	22) 46
7) 23	15) 5	23) 15
8) 105	16) −61	24) 50

Multiplication and Division

1) 1,680	11) 7,100	21) −29
2) 2,600	12) 26	22) 200
3) 140	13) −825	23) 0
4) −252	14) 4	24) −170
5) −121	15) −16	25) 9
6) −320	16) 125	26) −630
7) −96	17) −42	27) 3,634
8) 12,000	18) 1,032	28) 30
9) 1,200	19) −36	
10) 575	20) 840	

Ordering Integers and Numbers

1) -10, −5, −1, 2, 16	7) 29, 16, −4, −21, −25
2) −14, −8, 4, 11, 15	8) 71, 36, 12, −26, −54
3) −28, −21, −16, 37, 39	9) 55, 37, −17, −19, −46
4) −55, −27, −15, 35, 48	10) 95, 87, 37, −22, −46
5) −35, −32, 28, 42, 47	11) 84, 79, −9, −65, −78
6) −59, −24, 38, 85, 95	12) 39, 19, −45, −60, −80

Order of Operations

1) 18	2) 7	3) 75	4) −5

5) 38   9) 94   13) −9   17) 12

6) 24   10) 20   14) 20   18) 81

7) −27   11) 64   15) 576   19) 74

8) 72   12) 28   16) 32   20) −11

Factoring

1) 1, 3, 5, 15

2) 1, 41

3) 1, 2, 4, 8, 16, 32

4) 1, 5, 17, 85

5) 1, 2, 31, 62

6) 1, 5, 13, 65

7) 1, 2, 19, 38

8) 1, 2, 5, 10

9) 1, 2, 3, 6, 9, 18, 27, 54

10) 1, 11, 121

11) 1, 3, 5, 9, 15, 45

12) 1, 2, 3, 5, 6, 9, 10, 15, 18, 30, 45, 90

13) 1, 2, 3, 4, 6, 8, 12, 24

14) 1, 3, 11, 33

15) 1, 2, 4, 23, 46, 92

16) 1, 3, 19, 57

17) 1, 2, 43, 86

18) 1, 2, 4, 5, 8, 10, 20, 40

19) 1, 3, 5, 7, 15, 21, 35, 105

20) 1, 2, 4, 5, 8, 10, 16, 20, 40, 80

21) 1, 3, 9, 27, 81

22) 1, 2, 3, 4, 6, 9, 12, 18, 36

23) 1, 2, 3, 6, 7, 9, 14, 18, 21, 42, 63, 126

24) 1, 2, 5, 10, 11, 22, 55, 110

25) 1, 2, 17, 34

26) 1, 2, 7, 14, 49, 98

27) 1, 2, 3, 4, 6, 8, 12, 18, 24, 36, 72

28) 1, 2, 4, 8, 13, 26, 52, 104

Great Common Factor (GCF)

1) 2   9) 9   17) 8

2) 1   10) 9   18) 20

3) 5   11) 6   19) 1

4) 5   12) 35   20) 8

5) 4   13) 20   21) 10

6) 2   14) 2   22) 1

7) 5   15) 48   23) 30

8) 5   16) 5   24) 3

Least Common Multiple (LCM)

1) 80   3) 60   5) 30

2) 64   4) 84   6) 25

7) 240

8) 90

9) 702

10) 140

11) 54

12) 220

13) 120

14) 252

15) 40

16) 54

17) 38

18) 44

19) 300

20) 48

21) 340

22) 168

23) 140

24) 126

Divisibility Rule

1) 16   <u>2</u>, 3, <u>4</u>, 5, 6, 9, 10

2) 121   2, 3, 4, 5, 6, 9, 10

3) 252   <u>2</u>, <u>3</u>, <u>4</u>, 5, <u>6</u>, 9, 10

4) 74   <u>2</u>, 3, 4, 5, 6, 9, 10

5) 241   2, 3, 4, 5, 6, 9, 10

6) 155   2, 3, 4, <u>5</u>, 6, 9, 10

7) 65   2, 3, 4, <u>5</u>, 6, 9, 10

8) 320   <u>2</u>, 3, <u>4</u>, <u>5</u>, 6, 9, 10

9) 1,134   <u>2</u>, <u>3</u>, 4, 5, <u>6</u>, <u>9</u>, 10

10) 68   <u>2</u>, 3, <u>4</u>, 5, 6, 9, 10

11) 754   <u>2</u>, 3, 4, 5, 6, 9, 10

12) 128   <u>2</u>, 3, <u>4</u>, 5, 6, 9, 10

13) 38   <u>2</u>, 3, 4, 5, 6, 9, 10

14) 185   2, 3, 4, <u>5</u>, 6, 9, 10

15) 905   2, 3, 4, <u>5</u>, 6, 9, 10

16) 157   2, 3, 4, 5, 6, 9, 10

17) 540   <u>2</u>, <u>3</u>, <u>4</u>, <u>5</u>, <u>6</u>, <u>9</u>, <u>10</u>

18) 340   <u>2</u>, 3, <u>4</u>, <u>5</u>, 6, 9, <u>10</u>

19) 480   <u>2</u>, <u>3</u>, <u>4</u>, <u>5</u>, <u>6</u>, 9, <u>10</u>

20) 2,750   <u>2</u>, 3, 4, <u>5</u>, 6, 9, <u>10</u>

21) 330   <u>2</u>, <u>3</u>, 4, <u>5</u>, <u>6</u>, 9, <u>10</u>

22) 346   <u>2</u>, 3, 4, 5, 6, 9, 10

23) 108   <u>2</u>, <u>3</u>, <u>4</u>, 5, <u>6</u>, 9, 10

24) 656   <u>2</u>, 3, <u>4</u>, 5, 6, 9, 10

Chapter 2 :

Fractions

Adding Fractions – Unlike Denominator

Add the fractions and simplify the answers.

1) $\frac{1}{8} + \frac{2}{5} =$

2) $\frac{3}{7} + \frac{1}{2} =$

3) $\frac{1}{4} + \frac{2}{9} =$

4) $\frac{3}{5} + \frac{1}{2} =$

5) $\frac{7}{18} + \frac{1}{3} =$

6) $\frac{13}{54} + \frac{5}{18} =$

7) $\frac{5}{8} + \frac{1}{6} =$

8) $\frac{3}{10} + \frac{1}{4} =$

9) $\frac{5}{11} + \frac{2}{4} =$

10) $\frac{1}{9} + \frac{4}{7} =$

11) $\frac{5}{18} + \frac{3}{8} =$

12) $\frac{7}{32} + \frac{3}{4} =$

13) $\frac{18}{130} + \frac{4}{10} =$

14) $\frac{8}{63} + \frac{3}{7} =$

15) $\frac{11}{64} + \frac{1}{4} =$

16) $\frac{4}{15} + \frac{2}{5} =$

17) $\frac{4}{7} + \frac{3}{6} =$

18) $\frac{5}{72} + \frac{2}{9} =$

19) $\frac{2}{15} + \frac{1}{25} =$

20) $\frac{5}{12} + \frac{3}{8} =$

21) $\frac{7}{88} + \frac{1}{8} =$

22) $\frac{7}{12} + \frac{2}{5} =$

23) $\frac{3}{72} + \frac{1}{4} =$

24) $\frac{2}{27} + \frac{1}{18} =$

Subtracting Fractions – Unlike Denominator

Solve each problem.

1) $\frac{3}{5} - \frac{1}{6} =$

2) $\frac{5}{6} - \frac{1}{8} =$

3) $\frac{7}{6} - \frac{3}{11} =$

4) $\frac{5}{7} - \frac{4}{15} =$

5) $\frac{6}{7} - \frac{3}{14} =$

6) $\frac{7}{12} - \frac{7}{18} =$

7) $\frac{17}{20} - \frac{2}{5} =$

8) $\frac{2}{3} - \frac{1}{16} =$

9) $\frac{6}{7} - \frac{4}{9} =$

10) $\frac{3}{8} - \frac{5}{32} =$

11) $\frac{3}{5} - \frac{7}{40} =$

12) $\frac{5}{6} - \frac{7}{30} =$

13) $\frac{6}{7} - \frac{4}{21} =$

14) $\frac{5}{3} - \frac{8}{15} =$

15) $\frac{2}{11} - \frac{3}{22} =$

16) $\frac{5}{6} - \frac{4}{54} =$

17) $\frac{7}{24} - \frac{7}{32} =$

18) $\frac{3}{4} - \frac{3}{5} =$

19) $\frac{1}{2} - \frac{2}{9} =$

20) $\frac{5}{11} - \frac{3}{13} =$

Converting Mix Numbers

Convert the following mixed numbers into improper fractions.

1) $5\frac{5}{4} =$

2) $5\frac{7}{12} =$

3) $4\frac{1}{7} =$

4) $6\frac{1}{10} =$

5) $3\frac{1}{4} =$

6) $3\frac{19}{21} =$

7) $5\frac{9}{10} =$

8) $4\frac{7}{12} =$

9) $3\frac{10}{11} =$

10) $6\frac{2}{5} =$

11) $8\frac{2}{3} =$

12) $3\frac{7}{16} =$

13) $6\frac{8}{13} =$

14) $4\frac{8}{11} =$

15) $7\frac{1}{4} =$

16) $5\frac{6}{11} =$

17) $8\frac{1}{5} =$

18) $3\frac{7}{12} =$

19) $6\frac{1}{22} =$

20) $3\frac{2}{3} =$

21) $7\frac{1}{19} =$

22) $4\frac{7}{11} =$

23) $1\frac{5}{8} =$

24) $9\frac{5}{17} =$

Converting improper Fractions

Convert the following improper fractions into mixed numbers

1) $\frac{68}{19} =$

2) $\frac{79}{33} =$

3) $\frac{49}{17} =$

4) $\frac{56}{23} =$

5) $\frac{79}{18} =$

6) $\frac{137}{42} =$

7) $\frac{120}{33} =$

8) $\frac{26}{5} =$

9) $\frac{33}{19} =$

10) $\frac{13}{2} =$

11) $\frac{39}{4} =$

12) $\frac{161}{50} =$

13) $\frac{89}{77} =$

14) $\frac{42}{19} =$

15) $\frac{110}{13} =$

16) $\frac{65}{4} =$

17) $\frac{122}{9} =$

18) $\frac{81}{16} =$

19) $\frac{37}{6} =$

20) $\frac{67}{22} =$

21) $\frac{5}{4} =$

22) $\frac{79}{13} =$

23) $\frac{51}{11} =$

24) $\frac{36}{5} =$

Addition Mix Numbers

Add the following fractions.

1) $2\frac{1}{5} + 3\frac{2}{5} =$

2) $5\frac{3}{11} + 3\frac{4}{11} =$

3) $2\frac{2}{10} + 3\frac{1}{10} =$

4) $3\frac{3}{16} + 2\frac{1}{4} =$

5) $2\frac{3}{7} + 3\frac{4}{21} =$

6) $6\frac{2}{7} + 3\frac{1}{2} =$

7) $2\frac{8}{27} + 2\frac{2}{18} =$

8) $2\frac{3}{4} + 3\frac{1}{3} =$

9) $4\frac{5}{6} + 1\frac{1}{6} =$

10) $3\frac{5}{7} + 1\frac{3}{7} =$

11) $4\frac{1}{2} + 2\frac{2}{5} =$

12) $5\frac{1}{4} + 2\frac{5}{6} =$

13) $4\frac{1}{7} + 2\frac{2}{7} =$

14) $6\frac{5}{9} + 4\frac{2}{18} =$

15) $6\frac{3}{7} + 3\frac{1}{2} =$

16) $5\frac{2}{3} + 1\frac{4}{7} =$

17) $4\frac{5}{6} + 6\frac{1}{4} =$

18) $2\frac{2}{5} + 3\frac{3}{8} =$

19) $3\frac{1}{6} + 2\frac{4}{9} =$

20) $5\frac{3}{5} + 7\frac{2}{7} =$

21) $4\frac{5}{8} + 1\frac{1}{5} =$

22) $6\frac{1}{7} + 4\frac{4}{13} =$

23) $2\frac{7}{11} + 2\frac{4}{5} =$

24) $3\frac{1}{6} + 1\frac{5}{12} =$

Subtracting Mix Numbers

Subtract the following fractions.

1) $5\frac{1}{7} - 3\frac{1}{7} =$

2) $8\frac{5}{16} - 8\frac{2}{16} =$

3) $8\frac{5}{36} - 7\frac{1}{36} =$

4) $4\frac{1}{20} - 1\frac{1}{15} =$

5) $3\frac{1}{3} - 2\frac{1}{6} =$

6) $8\frac{1}{2} - 3\frac{2}{5} =$

7) $7\frac{5}{8} - 3\frac{3}{8} =$

8) $9\frac{9}{13} - 4\frac{6}{13} =$

9) $5\frac{7}{12} - 2\frac{5}{12} =$

10) $4\frac{4}{7} - 1\frac{3}{7} =$

11) $7\frac{1}{5} - 2\frac{1}{10} =$

12) $4\frac{5}{6} - 2\frac{1}{6} =$

13) $3\frac{2}{40} - 1\frac{1}{5} =$

14) $4\frac{1}{8} - 2\frac{1}{16} =$

15) $14\frac{4}{15} - 11\frac{2}{15} =$

16) $6\frac{2}{4} - 1\frac{1}{4} =$

17) $4\frac{1}{7} - 2\frac{3}{7} =$

18) $5\frac{1}{16} - 2\frac{1}{4} =$

19) $6\frac{2}{3} - 1\frac{1}{9} =$

20) $4\frac{3}{25} - 4\frac{1}{75} =$

21) $9\frac{9}{22} - 5\frac{1}{4} =$

22) $8\frac{4}{10} - 2\frac{3}{40} =$

23) $3\frac{2}{6} - 2\frac{1}{18} =$

24) $7\frac{9}{13} - 3\frac{3}{13} =$

Simplify Fractions

Reduce these fractions to lowest terms

1) $\frac{18}{12} =$

2) $\frac{10}{15} =$

3) $\frac{32}{40} =$

4) $\frac{27}{36} =$

5) $\frac{6}{36} =$

6) $\frac{27}{63} =$

7) $\frac{16}{28} =$

8) $\frac{48}{60} =$

9) $\frac{8}{72} =$

10) $\frac{30}{12} =$

11) $\frac{45}{60} =$

12) $\frac{30}{90} =$

13) $\frac{21}{35} =$

14) $\frac{7}{28} =$

15) $\frac{24}{84} =$

16) $\frac{34}{51} =$

17) $\frac{66}{55} =$

18) $\frac{36}{135} =$

19) $\frac{21}{56} =$

20) $\frac{64}{56} =$

21) $\frac{140}{280} =$

22) $\frac{138}{731} =$

23) $\frac{175}{35} =$

24) $\frac{170}{680} =$

Multiplying Fractions

Find the product.

1) $\frac{8}{5} \times \frac{2}{12} =$

2) $\frac{8}{44} \times \frac{10}{16} =$

3) $\frac{8}{30} \times \frac{12}{16} =$

4) $\frac{9}{14} \times \frac{21}{36} =$

5) $\frac{14}{15} \times \frac{5}{7} =$

6) $\frac{16}{19} \times \frac{3}{4} =$

7) $\frac{4}{9} \times \frac{9}{8} =$

8) $\frac{87}{63} \times 0 =$

9) $\frac{5}{16} \times \frac{32}{14} =$

10) $\frac{32}{45} \times \frac{5}{8} =$

11) $\frac{34}{26} \times \frac{13}{17} =$

12) $\frac{6}{42} \times \frac{7}{36} =$

13) $\frac{26}{16} \times \frac{24}{8} =$

14) $\frac{10}{27} \times \frac{18}{5} =$

15) $\frac{30}{54} \times \frac{16}{6} =$

16) $\frac{24}{14} \times 7 =$

17) $\frac{10}{33} \times \frac{66}{35} =$

18) $\frac{10}{18} \times \frac{9}{20} =$

19) $\frac{7}{11} \times \frac{8}{21} =$

20) $\frac{26}{24} \times \frac{8}{52} =$

21) $\frac{30}{15} \times \frac{1}{60} =$

22) $\frac{20}{27} \times \frac{18}{100} =$

23) $\frac{8}{21} \times \frac{7}{64} =$

24) $\frac{100}{200} \times \frac{600}{800} =$

Multiplying Mixed Number

Multiply. Reduce to lowest terms.

1) $2\frac{6}{10} \times 1\frac{6}{8} =$

2) $1\frac{15}{12} \times 1\frac{2}{9} =$

3) $2\frac{3}{7} \times 1\frac{2}{9} =$

4) $3\frac{1}{7} \times 2\frac{1}{2} =$

5) $4\frac{3}{4} \times 1\frac{1}{4} =$

6) $3\frac{1}{2} \times 1\frac{4}{5} =$

7) $3\frac{3}{4} \times 1\frac{1}{2} =$

8) $5\frac{2}{3} \times 3\frac{1}{3} =$

9) $3\frac{2}{3} \times 3\frac{1}{2} =$

10) $2\frac{1}{3} \times 3\frac{1}{2} =$

11) $4\frac{3}{4} \times 3\frac{2}{3} =$

12) $2\frac{4}{11} \times 2\frac{1}{7} =$

13) $2\frac{2}{7} \times 1\frac{1}{5} =$

14) $3\frac{1}{3} \times 1\frac{1}{5} =$

15) $2\frac{2}{3} \times 3\frac{1}{2} =$

16) $2\frac{1}{8} \times 2\frac{2}{5} =$

17) $2\frac{1}{4} \times 1\frac{2}{3} =$

18) $2\frac{3}{5} \times 1\frac{1}{4} =$

19) $2\frac{3}{5} \times 1\frac{5}{8} =$

20) $3\frac{1}{6} \times 2\frac{5}{7} =$

21) $2\frac{5}{8} \times 1\frac{1}{5} =$

22) $2\frac{5}{7} \times 3\frac{1}{6} =$

Dividing Fractions

Divide these fractions.

1) $1 \div \frac{1}{8} =$

2) $\frac{9}{17} \div 9 =$

3) $\frac{11}{20} \div \frac{5}{11} =$

4) $\frac{25}{60} \div \frac{5}{4} =$

5) $\frac{6}{23} \div \frac{4}{23} =$

6) $\frac{4}{16} \div \frac{18}{24} =$

7) $0 \div \frac{1}{9} =$

8) $\frac{12}{16} \div \frac{8}{9} =$

9) $\frac{8}{12} \div \frac{4}{18} =$

10) $\frac{9}{14} \div \frac{3}{7} =$

11) $\frac{8}{15} \div \frac{25}{16} =$

12) $\frac{35}{16} \div \frac{15}{8} =$

13) $\frac{11}{15} \div \frac{11}{5} =$

14) $\frac{8}{16} \div \frac{20}{6} =$

15) $\frac{40}{24} \div \frac{48}{80} =$

16) $\frac{7}{30} \div \frac{63}{5} =$

17) $\frac{36}{8} \div \frac{18}{24} =$

18) $9 \div \frac{1}{2} =$

19) $\frac{48}{35} \div \frac{8}{7} =$

20) $\frac{3}{36} \div \frac{9}{6} =$

21) $\frac{4}{7} \div \frac{12}{14} =$

22) $\frac{8}{40} \div \frac{10}{5} =$

Dividing Mixed Number

Divide the following mixed numbers. Cancel and simplify when possible.

1) $4\frac{1}{6} \div 4\frac{1}{5} =$

2) $3\frac{1}{8} \div 1\frac{1}{4} =$

3) $3\frac{1}{4} \div 2\frac{2}{7} =$

4) $4\frac{1}{3} \div 4\frac{1}{2} =$

5) $3\frac{1}{7} \div 1\frac{2}{5} =$

6) $3\frac{3}{5} \div 2\frac{2}{6} =$

7) $4\frac{3}{5} \div 2\frac{1}{3} =$

8) $2\frac{4}{9} \div 1\frac{1}{9} =$

9) $3\frac{5}{6} \div 3\frac{1}{2} =$

10) $9\frac{1}{9} \div 3\frac{2}{3} =$

11) $2\frac{2}{7} \div 4\frac{1}{7} =$

12) $4\frac{3}{8} \div 1\frac{3}{4} =$

13) $5\frac{1}{8} \div 1\frac{1}{12} =$

14) $6\frac{3}{8} \div 3\frac{1}{3} =$

15) $4\frac{2}{5} \div 1\frac{1}{5} =$

16) $2\frac{1}{2} \div 2\frac{2}{9} =$

17) $7\frac{1}{6} \div 5\frac{3}{8} =$

18) $5\frac{1}{2} \div 4\frac{1}{3} =$

19) $4\frac{5}{7} \div 1\frac{1}{3} =$

20) $3\frac{5}{6} \div 1\frac{1}{4} =$

21) $8\frac{1}{3} \div 5\frac{1}{4} =$

22) $3\frac{1}{11} \div 1\frac{1}{5} =$

23) $4\frac{1}{6} \div 5\frac{5}{6} =$

24) $2\frac{1}{14} \div 2\frac{1}{7} =$

Comparing Fractions

Compare the fractions, and write >, < or =

1) $\dfrac{28}{3}$ _____ $\dfrac{48}{15}$

2) $\dfrac{96}{3}$ _____ $\dfrac{14}{5}$

3) $\dfrac{8}{9}$ _____ $\dfrac{6}{4}$

4) $\dfrac{12}{4}$ _____ $\dfrac{13}{9}$

5) $\dfrac{1}{8}$ _____ $\dfrac{2}{3}$

6) $\dfrac{10}{6}$ _____ $\dfrac{16}{7}$

7) $\dfrac{12}{13}$ _____ $\dfrac{7}{9}$

8) $\dfrac{20}{14}$ _____ $\dfrac{25}{3}$

9) $4\dfrac{1}{12}$ _____ $6\dfrac{1}{3}$

10) $8\dfrac{1}{6}$ _____ $3\dfrac{1}{8}$

11) $3\dfrac{1}{2}$ _____ $3\dfrac{1}{5}$

12) $7\dfrac{5}{8}$ _____ $7\dfrac{2}{9}$

13) $3\dfrac{4}{16}$ _____ $5\dfrac{6}{10}$

14) $\dfrac{1}{15}$ _____ $\dfrac{3}{7}$

15) $\dfrac{31}{25}$ _____ $\dfrac{19}{83}$

16) $\dfrac{12}{100}$ _____ $\dfrac{6}{62}$

17) $15\dfrac{1}{4}$ _____ $15\dfrac{1}{9}$

18) $\dfrac{1}{5}$ _____ $\dfrac{1}{9}$

19) $\dfrac{1}{7}$ _____ $\dfrac{1}{13}$

20) $\dfrac{1}{18}$ _____ $\dfrac{8}{15}$

21) $\dfrac{7}{22}$ _____ $\dfrac{9}{76}$

22) $\dfrac{4}{5}$ _____ $\dfrac{2}{5}$

23) $3\dfrac{7}{6}$ _____ $4\dfrac{2}{12}$

24) $3\dfrac{25}{6}$ _____ $4\dfrac{5}{6}$

Answers of Worksheets

Adding Fractions – Unlike Denominator

1) $\frac{21}{40}$

2) $\frac{13}{14}$

3) $\frac{17}{36}$

4) $\frac{27}{22}$

5) $\frac{13}{18}$

6) $\frac{14}{27}$

7) $\frac{19}{24}$

8) $\frac{11}{20}$

9) $\frac{21}{22}$

10) $\frac{43}{63}$

11) $\frac{47}{72}$

12) $\frac{31}{32}$

13) $\frac{7}{12}$

14) $\frac{5}{9}$

15) $\frac{27}{64}$

16) $\frac{2}{3}$

17) $\frac{15}{14}$

18) $\frac{7}{24}$

19) $\frac{13}{75}$

20) $\frac{19}{24}$

21) $\frac{9}{44}$

22) $\frac{59}{60}$

23) $\frac{7}{24}$

24) $\frac{7}{54}$

Subtracting Fractions – Unlike Denominator

1) $\frac{13}{30}$

2) $\frac{17}{24}$

3) $\frac{59}{66}$

4) $\frac{47}{105}$

5) $\frac{9}{14}$

6) $\frac{7}{36}$

7) $\frac{9}{20}$

8) $\frac{29}{48}$

9) $\frac{26}{63}$

10) $\frac{7}{32}$

11) $\frac{17}{40}$

12) $\frac{3}{5}$

13) $\frac{2}{3}$

14) $\frac{17}{15}$

15) $\frac{1}{22}$

16) $\frac{41}{54}$

17) $\frac{7}{96}$

18) $\frac{3}{20}$

19) $\frac{5}{18}$

20) $\frac{22}{143}$

Converting Mix Numbers

1) $\frac{25}{4}$

2) $\frac{67}{12}$

3) $\frac{29}{7}$

4) $\frac{61}{10}$

5) $\frac{13}{4}$

6) $\frac{82}{21}$

7) $\frac{59}{10}$

8) $\frac{55}{12}$

9) $\frac{43}{11}$

10) $\frac{32}{5}$

11) $\frac{26}{3}$

12) $\frac{55}{16}$

13) $\frac{86}{13}$

14) $\frac{52}{11}$

15) $\frac{29}{4}$

16) $\frac{61}{11}$

17) $\frac{41}{5}$

18) $\frac{43}{12}$

19) $\frac{133}{22}$

20) $\frac{11}{3}$

21) $\frac{134}{19}$

22) $\frac{51}{11}$

23) $\frac{13}{8}$

24) $\frac{158}{17}$

Converting improper Fractions

1) $3\frac{11}{19}$

2) $2\frac{13}{33}$

3) $2\frac{15}{17}$

4) $2\frac{10}{23}$

5) $4\frac{7}{18}$

6) $3\frac{11}{42}$

7) $3\frac{21}{33}$

8) $5\frac{1}{5}$

9) $1\frac{14}{19}$

10) $6\frac{1}{2}$

11) $9\frac{3}{4}$

12) $3\frac{11}{50}$

13) $1\frac{12}{77}$

14) $2\frac{4}{19}$

15) $8\frac{6}{13}$

16) $16\frac{1}{4}$

17) $13\frac{5}{9}$

18) $5\frac{1}{16}$

19) $6\frac{1}{6}$

20) $3\frac{1}{22}$

21) $1\frac{1}{4}$

22) $6\frac{1}{13}$

23) $5\frac{1}{11}$

24) $7\frac{1}{5}$

Adding Mix Numbers

1) $10\frac{3}{5}$

2) $8\frac{7}{11}$

3) $5\frac{3}{10}$

4) $7\frac{5}{16}$

5) $4\frac{19}{21}$

6) $24\frac{5}{14}$

7) $4\frac{11}{27}$

8) $6\frac{1}{12}$

9) 9

10) $5\frac{1}{7}$

11) $6\frac{9}{10}$

12) $8\frac{1}{12}$

13) 8

14) $11\frac{1}{3}$

15) $24\frac{1}{2}$

16) $7\frac{5}{21}$

17) $11\frac{1}{12}$

18) $5\frac{31}{40}$

19) $5\frac{11}{18}$

20) $20\frac{31}{35}$

21) $7\frac{33}{40}$

22) $12\frac{9}{91}$

23) $7\frac{14}{55}$

24) $6\frac{5}{12}$

Subtracting Mix Numbers

1) 2

2) $\frac{3}{16}$

3) $1\frac{1}{9}$

4) $2\frac{59}{60}$

5) $1\frac{1}{6}$

6) $5\frac{1}{10}$

7) $4\frac{1}{4}$

8) $5\frac{3}{13}$

9) $3\frac{1}{6}$

10) $3\frac{1}{7}$

11) $5\frac{1}{10}$

12) $2\frac{2}{3}$

13) $1\frac{17}{20}$

14) $2\frac{1}{16}$

15) $3\frac{2}{15}$

16) $5\frac{1}{4}$

17) $1\frac{5}{7}$

18) $2\frac{13}{16}$

19) $5\frac{5}{9}$

20) $\frac{8}{75}$

21) $4\frac{7}{44}$

22) $6\frac{13}{40}$

23) $1\frac{5}{18}$

24) $4\frac{6}{13}$

Simplify Fractions

1) $\frac{3}{2}$

2) $\frac{2}{3}$

3) $\frac{4}{5}$

4) $\frac{3}{4}$

5) $\frac{1}{6}$

6) $\frac{3}{7}$

7) $\frac{4}{7}$

8) $\frac{4}{5}$

9) $\frac{1}{9}$

10) $\frac{5}{2}$

11) $\frac{3}{4}$

12) $\frac{1}{3}$

13) $\frac{3}{5}$

14) $\frac{1}{4}$

15) $\frac{2}{7}$

16) $\frac{2}{3}$

17) $\frac{6}{5}$

18) $\frac{4}{15}$

19) $\frac{3}{8}$

20) $\frac{8}{7}$

21) $\frac{1}{2}$

22) $\frac{6}{31}$

23) 5

24) $\frac{1}{4}$

Multiplying Fractions

1) $\frac{4}{15}$

2) $\frac{5}{44}$

3) $\frac{1}{5}$

4) $\frac{3}{8}$

5) $\frac{2}{3}$

6) $\frac{12}{19}$

7) $\frac{1}{2}$

8) 0

9) $\frac{5}{7}$

10) $\frac{4}{9}$

11) 1

12) $\frac{1}{36}$

13) $\frac{39}{8}$

14) $\frac{4}{3}$

15) $\frac{40}{27}$

16) 12

17) $\frac{4}{7}$

18) $\frac{1}{4}$

19) $\frac{8}{33}$

20) $\frac{1}{6}$

21) $\frac{1}{30}$

22) $\frac{2}{15}$

23) $\frac{1}{24}$

24) $\frac{3}{8}$

Multiplying Mixed Number

1) $4\frac{11}{20}$

2) $2\frac{3}{4}$

3) $2\frac{61}{63}$

4) $7\frac{6}{7}$

5) $5\frac{15}{16}$

6) $6\frac{3}{10}$

7) $5\frac{5}{8}$

8) $18\frac{8}{9}$

9) $12\frac{5}{6}$

10) $8\frac{1}{6}$

11) $17\frac{5}{12}$

12) $5\frac{5}{77}$

13) $2\frac{26}{35}$

14) 4

15) $9\frac{1}{3}$

16) $5\frac{1}{10}$

17) $3\frac{3}{4}$

18) $3\frac{1}{4}$

19) $4\frac{9}{40}$

20) $8\frac{25}{42}$

21) $3\frac{3}{20}$

22) $8\frac{25}{42}$

Dividing Fractions

1) 8

2) $\frac{1}{17}$

3) $\frac{1}{100}$

4) $\frac{1}{48}$

5) $\frac{3}{2}$

6) $\frac{1}{3}$

7) 0

8) $\frac{27}{32}$

9) 3

10) $\frac{3}{2}$

11) $\frac{128}{375}$

12) $\frac{7}{6}$

13) $\frac{1}{75}$

14) $\frac{1}{240}$

15) 1

16) $\frac{1}{54}$

17) 6

18) 18

19) $\frac{6}{5}$

20) $\frac{1}{18}$

21) $\frac{2}{3}$

22) $\frac{1}{10}$

Dividing Mixed Number

1) $\frac{125}{126}$

2) $\frac{5}{2}$

3) $1\frac{27}{64}$

4) $\frac{26}{27}$

5) $2\frac{12}{49}$

6) $1\frac{19}{35}$

7) $1\frac{34}{35}$

8) $\frac{2}{81}$

9) $1\frac{2}{21}$

10) $2\frac{16}{33}$

11) $\frac{16}{29}$

12) $2\frac{1}{2}$

13) $4\frac{19}{26}$

14) $1\frac{73}{80}$

15) $3\frac{2}{3}$

16) $1\frac{1}{8}$

17) $1\frac{1}{3}$

18) $1\frac{7}{26}$

19) $3\frac{15}{28}$

20) $3\frac{1}{15}$

21) $1\frac{37}{63}$

22) $2\frac{19}{33}$

23) $\frac{5}{7}$

24) $\frac{29}{30}$

Comparing Fractions

1) >	7) >	13) <	19) >
2) >	8) <	14) <	20) <
3) <	9) <	15) >	21) >
4) >	10) >	16) <	22) >
5) <	11) >	17) >	23) =
6) <	12) >	18) >	24) >

Chapter 3 :

Decimal

Round Decimals

Round each number to the correct place value

1) 0.<u>7</u>3 =

2) 5.<u>0</u>2 =

3) 10.<u>7</u>11 =

4) 0.<u>4</u>67 =

5) <u>8</u>.924 =

6) 0.0<u>7</u>5 =

7) 8.<u>1</u>2 =

8) 63.7<u>4</u>0 =

9) 2.5<u>3</u>8 =

10) 12.<u>2</u>97 =

11) 2.<u>0</u>8 =

12) 5.<u>3</u>24 =

13) 2.<u>1</u>32 =

14) 8.0<u>7</u>32 =

15) 7<u>5</u>.78 =

16) 4<u>8</u>.24 =

17) 6<u>2</u>7.132 =

18) 624.<u>7</u>88 =

19) 17.4<u>8</u>1 =

20) 9<u>4</u>.86 =

21) 4.3<u>0</u>67 =

22) 57.<u>0</u>86 =

23) 224.<u>2</u>24 =

24) 0.1<u>3</u>44 =

25) 0.00<u>7</u>8 =

26) 7.0<u>3</u>67 =

27) 15.4<u>4</u>33 =

28) 21.0<u>9</u>31 =

Decimals Addition

Add the following.

1) 34.21
   + 14.25
   _____

2) 0.66
   + 0.31
   _____

3) 25.36
   + 20.87
   _____

4) 75.165
   + 4.105
   _____

5) 8.650
   + 7.82
   _____

6) 5.324
   + 2.138
   _____

7) 81.21
   + 15.85
   _____

8) 71.05
   + 11.35
   _____

9) 26.21
   + 8.07
   _____

10) 6.96
   + 13.23
   _____

11) 15.214
   + 11.251
   _____

12) 72.36
   + 5.32
   _____

13) 52.05
   + 10.54
   _____

14) 107.11
   + 5.05
   _____

Decimals Subtraction

Subtract the following

1)
$$\begin{array}{r} 7.45 \\ -\ 5.12 \\ \hline \end{array}$$

8)
$$\begin{array}{r} 42.56 \\ -\ 22.45 \\ \hline \end{array}$$

2)
$$\begin{array}{r} 85.35 \\ -\ 72.37 \\ \hline \end{array}$$

9)
$$\begin{array}{r} 58.13 \\ -\ 32.35 \\ \hline \end{array}$$

3)
$$\begin{array}{r} 0.82 \\ -\ 0.6 \\ \hline \end{array}$$

10)
$$\begin{array}{r} 8.763 \\ -\ 0.425 \\ \hline \end{array}$$

4)
$$\begin{array}{r} 11.245 \\ -\ 8.6 \\ \hline \end{array}$$

11)
$$\begin{array}{r} 55.69 \\ -\ 45.32 \\ \hline \end{array}$$

5)
$$\begin{array}{r} 0.652 \\ -\ 0.09 \\ \hline \end{array}$$

12)
$$\begin{array}{r} 10.352 \\ -\ 4.325 \\ \hline \end{array}$$

6)
$$\begin{array}{r} 75.25 \\ -\ 28.88 \\ \hline \end{array}$$

13)
$$\begin{array}{r} 11.105 \\ -\ 3.128 \\ \hline \end{array}$$

7)
$$\begin{array}{r} 112.66 \\ -\ 88.98 \\ \hline \end{array}$$

14)
$$\begin{array}{r} 126.78 \\ -\ 8.52 \\ \hline \end{array}$$

Decimals Multiplication

Solve.

1) 2.1
   × 4.4
   _____

2) 5.2
   × 3.7
   _____

3) 7.04
   × 3.04
   _____

4) 55.02
   × 100
   _____

5) 61.8
   × 10
   _____

6) 35.62
   × 5.5
   _____

7) 32.75
   × 11.3
   _____

8) 1.65
   × 7.35
   _____

9) 10.05
   × 0.06
   _____

10) 21.04
   × 6.08
   _____

11) 10.34
   × 11.2
   _____

12) 7.67
   × 0.05
   _____

13) 7.2
   × 0.16
   _____

14) 13.2
   × 4.05
   _____

Decimal Division

Dividing Decimals.

1) $7 \div 10,000 =$

2) $6 \div 100 =$

3) $7.1 \div 100 =$

4) $0.004 \div 10 =$

5) $9 \div 81 =$

6) $8 \div 64 =$

7) $5 \div 45 =$

8) $9 \div 180 =$

9) $7 \div 1,000 =$

10) $0.6 \div 0.63 =$

11) $0.9 \div 0.009 =$

12) $0.6 \div 0.12 =$

13) $0.6 \div 0.42 =$

14) $0.4 \div 0.04 =$

15) $3.08 \div 10 =$

16) $9.4 \div 10 =$

17) $6.75 \div 100 =$

18) $18.3 \div 3.3 =$

19) $64.4 \div 4 =$

20) $0.4 \div 0.004 =$

21) $7.05 \div 3.5 =$

22) $0.08 \div 0.40 =$

23) $1.8 \div 15.2 =$

24) $0.18 \div 108 =$

25) $15.72 \div 1.5 =$

26) $0.05 \div 250 =$

Comparing Decimals

Write the Correct Comparison Symbol (>, < or =)

1) 2.62 ____ 3.62

2) 0.8 ____ 0.726

3) 15.6 ____ 15.600

4) 8.07 ____ 8.70

5) 0.922 ____ 0.92

6) 0.856 ____ 0.956

7) 4.34 ____ 4.242

8) 5.0025 ____ 5.025

9) 24.087 ____ 24.078

10) 7.12 ____ 7.29

11) 4.44 ____ 4.444

12) 0.09 ____ 0.18

13) 1.302 ____ 1.32

14) 9.56 ____ 9.0569

15) 0.55 ____ 0.055

16) 61.04 ____ 61.040

17) 0.350 ____ 0.45

18) 53.92 ____ 55.01

19) 0.075 ____ 0.705

20) 46.5 ____ 39.8

21) 7.89 ____ 10.2

22) 0.014 ____ 0.0104

23) 9.042 ____ 0.9042

24) 6.5 ____ 0.658

25) 8.5 ____ 0.859

26) 6.32 ____ 6.3200

27) 1.43 ____ 0.143

28) 7.0809 ____ 7.0890

Convert Fraction to Decimal

Write each as a decimal.

1) $\frac{25}{50} =$

2) $\frac{92}{200} =$

3) $\frac{24}{150} =$

4) $\frac{32}{64} =$

5) $\frac{8}{72} =$

6) $\frac{56}{100} =$

7) $\frac{4}{50} =$

8) $\frac{31}{48} =$

9) $\frac{27}{300} =$

10) $\frac{15}{55} =$

11) $\frac{16}{32} =$

12) $\frac{12}{32} =$

13) $\frac{6}{20} =$

14) $\frac{18}{250} =$

15) $\frac{24}{80} =$

16) $\frac{30}{40} =$

17) $\frac{68}{100} =$

18) $\frac{7}{35} =$

19) $\frac{87}{100} =$

20) $\frac{1}{120} =$

21) $\frac{30}{180} =$

22) $\frac{6}{240} =$

Convert Decimal to Percent

Write each as a percent.

1) $0.285 =$

2) $0.14 =$

3) $3.2 =$

4) $0.019 =$

5) $0.007 =$

6) $0.786 =$

7) $0.245 =$

8) $0.57 =$

9) $0.002 =$

10) $0.205 =$

11) $0.324 =$

12) $84.9 =$

13) $3.015 =$

14) $0.9 =$

15) $7.35 =$

16) $0.0312 =$

17) $0.0061 =$

18) $0.960 =$

19) $6.68 =$

20) $0.484 =$

21) $8.957 =$

22) $0.879 =$

23) $2.7 =$

24) $0.7 =$

25) $2.6 =$

26) $36.2 =$

27) $1.52 =$

28) $0.008 =$

Convert Fraction to Percent

Write each as a percent.

1) $\frac{2}{8} =$

2) $\frac{3}{8} =$

3) $\frac{14}{28} =$

4) $\frac{30}{70} =$

5) $\frac{12}{28} =$

6) $\frac{17}{68} =$

7) $\frac{8}{11} =$

8) $\frac{14}{30} =$

9) $\frac{6}{50} =$

10) $\frac{12}{48} =$

11) $\frac{5}{34} =$

12) $\frac{81}{30} =$

13) $\frac{48}{160} =$

14) $\frac{32}{50} =$

15) $\frac{16}{58} =$

16) $\frac{2}{22} =$

17) $\frac{32}{88} =$

18) $\frac{21}{36} =$

19) $\frac{18}{92} =$

20) $\frac{8}{80} =$

21) $\frac{72}{900} =$

22) $\frac{360}{180} =$

Answers of Worksheets

Round Decimals

1) 0.7	11) 2.1	21) 4.31
2) 5.0	12) 5.3	22) 57.1
3) 10.7	13) 2.1	23) 224.2
4) 0.5	14) 8.07	24) 0.13
5) 9.0	15) 76.0	25) 0.008
6) 0.08	16) 48.0	26) 7.04
7) 8.1	17) 630.0	27) 15.44
8) 63.74	18) 624.8	28) 21.09
9) 2.54	19) 17.48	
10) 12.3	20) 95.0	

Decimals Addition

1) 48.46	6) 7.462	11) 26.465
2) 0.97	7) 97.06	12) 77.68
3) 46.23	8) 82.4	13) 62.59
4) 79.27	9) 34.28	14) 112.16
5) 16.47	10) 20.19	

Decimals Subtraction

1) 2.33	6) 46.37	11) 10.37
2) 12.98	7) 23.68	12) 6.027
3) 0.22	8) 20.11	13) 7.977
4) 2.645	9) 25.78	14) 118.26
5) 0.562	10) 8.338	

Decimals Multiplication

1) 9.24	6) 195.91	11) 115.808
2) 19.24	7) 370.075	12) 0.3835
3) 21.4016	8) 12.1275	13) 1.152
4) 5,502	9) 0.603	14) 53.46
5) 618	10) 127.9232	

Decimal Division

1) 0.0007	2) 0.06	3) 0.071

4) 0.0004	12) 5	20) 100
5) 0.111….	13) 1.4285…	21) 2.01428…
6) 0.125	14) 10	22) 0.2
7) 0.111…	15) 0.308	23) 0.1184…
8) 0.05	16) 0.94	24) 0.0017
9) 0.007	17) 0.0675	25) 10.48
10) 0.952…	18) 5.5454…	26) 0.0002
11) 100	19) 16.1	

Comparing Decimals

1) <	11) <	21) <
2) >	12) <	22) >
3) =	13) <	23) >
4) <	14) >	24) >
5) >	15) >	25) >
6) <	16) =	26) =
7) >	17) <	27) >
8) <	18) <	28) <
9) >	19) <	
10) >	20) >	

Convert Fraction to Decimal

1) 0.5	9) 0.09	17) 0.68
2) 0.46	10) 0.27	18) 0.2
3) 0.16	11) 0.5	19) 0.87
4) 0.5	12) 0.375	20) 0.0083
5) 0.11	13) 0.3	21) 0.166
6) 0.56	14) 0.072	22) 0.025
7) 0.08	15) 0.3	
8) 0.646	16) 0.75	

Convert Decimal to Percent

1) 28.5%	4) 1.9%	7) 24.5%
2) 14%	5) 0.7%	8) 57%
3) 320%	6) 78.6%	9) 0.2%

10) 20.5%

11) 32.4%

12) 8,490%

13) 301.5%

14) 90%

15) 735%

16) 3.12%

17) 0.61%

18) 96%

19) 668%

20) 48.4%

21) 895.7%

22) 87.9%

23) 270%

24) 70%

25) 260%

26) 3,620%

27) 152%

28) 0.8%

Convert Fraction to Percent

1) 25%

2) 37.5%

3) 50%

4) 42.86%

5) 29.31%

6) 25%

7) 72.72%

8) 46.66%

9) 12%

10) 25%

11) 14.7%

12) 2.7%

13) 30%

14) 64%

15) 27.58%

16) 9.09%

17) 36.36%

18) 58.33%

19) 19.56%

20) 10%

21) 8%

22) 200%

Chapter 4 :

Geometry

Angles

What is the value of x in the following figures?

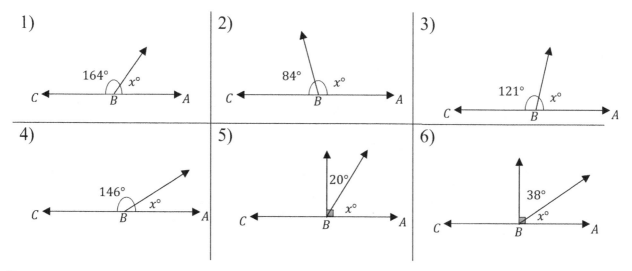

 Calculate.

7) Two supplement angles have equal measures. What is the measure of each angle? _____

8) The measure of an angle is seven fifth the measure of its supplement. What is the measure of the angle? _____

9) Two angles are complementary and the measure of one angle is 24 less than the other. What is the measure of the smaller angle?

_____

10) Two angles are complementary. The measure of one angle is one fifth the measure of the other. What is the measure of the bigger angle?

_____

11) Two supplementary angles are given. The measure of one angle is 40° less than the measure of the other. What does the smaller angle measure? _____

Area and Perimeter of Trapezoid

Find the perimeter and area of each trapezoid.

1)

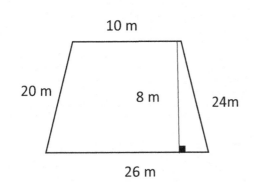

Perimeter:_____:

Area:_____:

2)

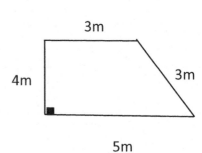

Perimeter:_____:

Area:_____:

3)

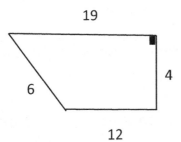

Perimeter:_____:

Area:_____:

4)

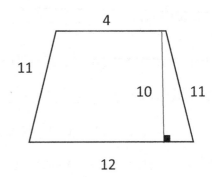

Perimeter:_____:

Area:_____:

5)

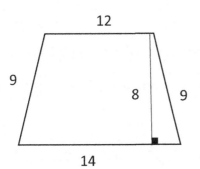

Perimeter:_____:

Area:_____:

6)

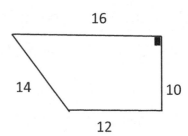

Perimeter:_____:

Area:_____:

Area and Perimeter of Parallelogram

Find the perimeter and area of each parallelogram.

1)

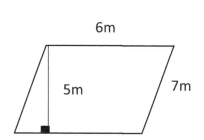

Perimeter:_____:

Area:_____:

2)

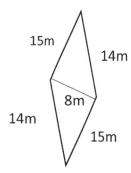

Perimeter:_____:

Area:_____:

3)

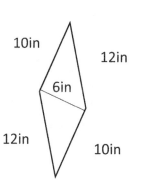

Perimeter:_____:

Area _____:

4)

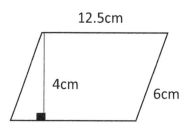

Perimeter:_____:

Area:_____:

5)

13m

12m

Perimeter:_____:

Area:_____:

6)

13m

Perimeter:_____:

Area:_____:

Circumference and Area of Circle

Find the circumference and area of each ($\pi = 3.14$).

1)

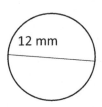

Circumference:

Area:

2)

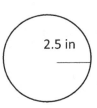

Circumference:_____.

Area:_____.

3)

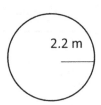

Circumference:_____.

Area _____.

4)

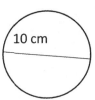

Circumference:_____.

Area:_____.

5)

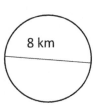

Circumference:_____.

Area:_____.

6)

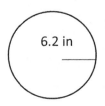

Circumference:_____.

Area:_____.

Perimeter of Polygon

Find the perimeter of each polygon.

1)

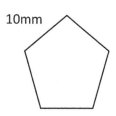

10mm

Perimeter:_____.

2)

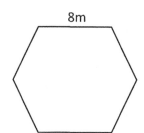

8m

Perimeter:_____:

3)

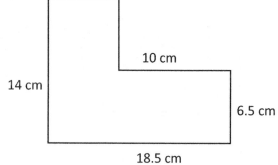

10 cm

14 cm

6.5 cm

18.5 cm

Perimeter:_____:

4)

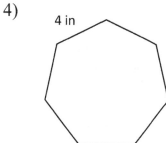

4 in

Perimeter:_____:

5)

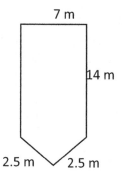

7 m

14 m

2.5 m   2.5 m

Perimeter:_____.

6)

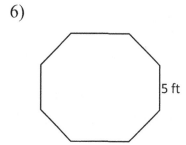

5 ft

Perimeter:_____:

Volume of Cubes

Find the volume of each cube.

1)

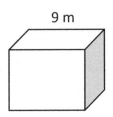

9 m

V:_____.

2)

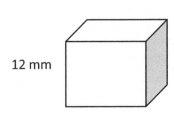

12 mm

V:_____.

3)

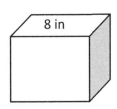

8 in

V:_____.

4)

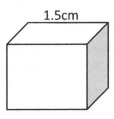

1.5cm

V:_____.

5)

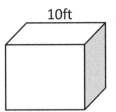

10ft

V:_____.

6)

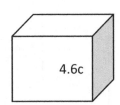

4.6c

V:_____.

Volume of Rectangle Prism

Find the volume of each rectangle prism.

1)

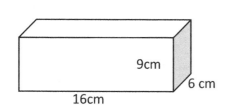

V:..:

2)

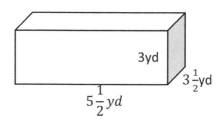

V:..:

3)

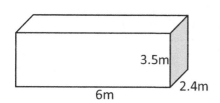

V:..:

4)

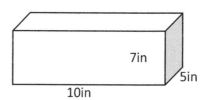

V:..:

5)

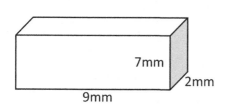

V:..:

6)

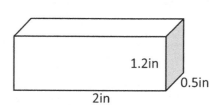

V:..:

Volume of Cylinder

Find the volume of each cylinder.

1)

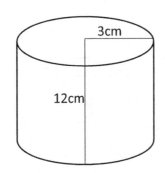

V:_____:

2)

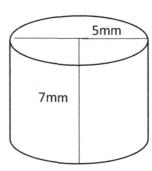

V:_____:

3)

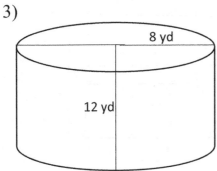

V:_____:

4)

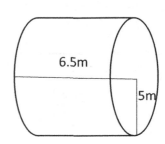

V:_____:

5)

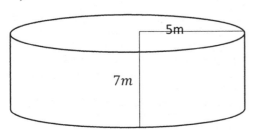

V:_____:

6)

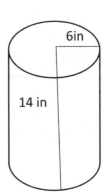

V:_____:

Volume of Pyramid and Cone

Find the volume of each pyramid and cone ($\pi = 3.14$).

1)

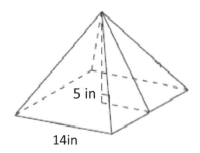

5 in

14in

V:......................................:

2)

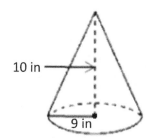

10 in

9 in

V:......................................:

3)

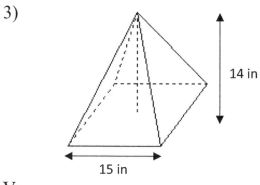

14 in

15 in

V:......................................:

4)

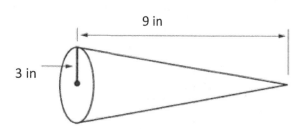

9 in

3 in

V:......................................:

5)

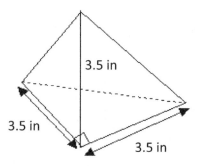

3.5 in

3.5 in

3.5 in

V:......................................:

6)

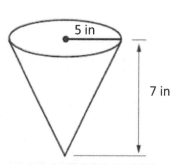

5 in

7 in

V:......................................:

Surface Area Cubes

Find the surface area of each cube.

1)

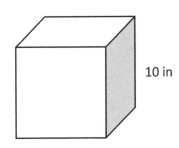

10 in

SA:_____:

2)

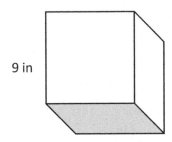

9 in

SA:_____:

3)

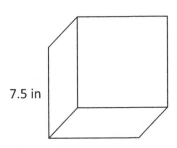

7.5 in

SA:_____:

4)

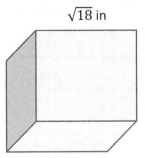

√18 in

SA:_____:

5)

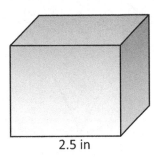

2.5 in

SA:_____:

6)

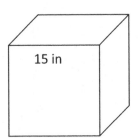

15 in

SA:_____:

Surface Area Rectangle Prism

Find the surface area of each rectangular prism.

1)

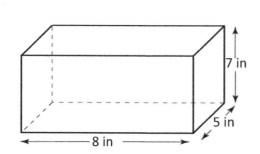

SA:_____:

2)

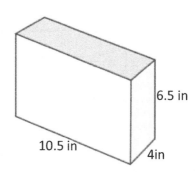

SA:_____:

3)

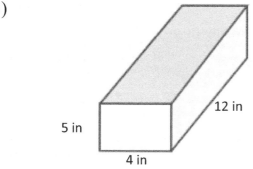

SA:_____:

4)

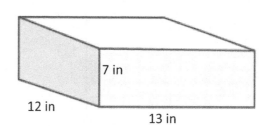

SA:_____:

5)

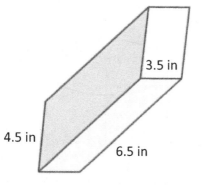

SA:_____:

6)

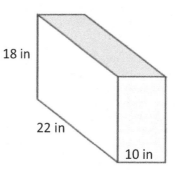

SA:_____:

Surface Area Cylinder

Find the surface area of each cylinder.

1)

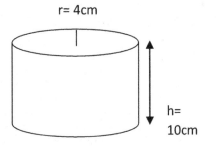

r= 4cm

h=
10cm

SA:_____:

2)

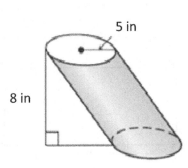

5 in

8 in

SA:_____:

3)

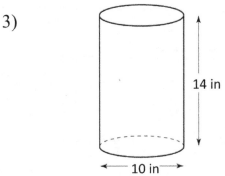

14 in

10 in

SA:_____:

4)

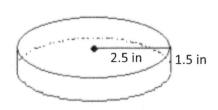

2.5 in   1.5 in

SA:_____:

5)

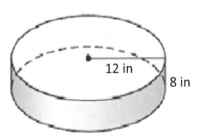

12 in

8 in

SA:_____:

6)

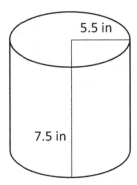

5.5 in

7.5 in

SA:_____:

Answers of Worksheets

Angles

1) 16°   4) 34°   7) 90°   10) 75°

2) 96°   5) 70°   8) 75°   11) 70°

3) 59°   6) 52°   9) 33°

Area and Perimeter of Trapezoid

1- Perimeter: 80,   Area:144   3- Perimeter: 41,   Area:62   5- Perimeter: 44,   Area:104

2- Perimeter: 15,   Area:16   4- Perimeter: 38,   Area:80   6- Perimeter: 52,   Area:140

Area and Perimeter of Parallelogram

1- Perimeter: $26m$, Area:$30(m)^2$ 4- Perimeter: $37cm$, Area:$50(cm)^2$

2- Perimeter: $58m$, Area:$120(m)^2$ 5- Perimeter: $50m$,   Area:$156(m)^2$

3- Perimeter: $44in$,   Area:$60(in)^2$ 6- Perimeter: $52m$, Area:$169(m)^2$

Circumference and Area of Circle

1) Circumference:37.68 mm   Area:$113.04(mm)^2$

2) Circumference: 15.7 in   Area:$19.625(in)^2$

3) Circumference: 13.816 m   Area:$15.197(m)^2$

4) Circumference: 31.4 cm   Area:$78.5(cm)^2$

5) Circumference: 25.12 in   Area:$50.24(in)^2$

6) Circumference: 38.936 km   Area:$120.702(km)^2$

Perimeter of Polygon

1) 50 mm   3) 65 cm   5) 40 m

2) 48 m   4) 28 in   6) 40 ft

Volume of Cubes

1) $729m^3$ 4) $3.375(cm)^3$

2) $1,728(mm)^3$ 5) $1,000(ft)^3$

3) $512in^3$ 6) $97.336(cm)^3$

Volume of Rectangle Prism

1) $864(cm)^3$ 3) $50.4(m)^3$ 5) $126(mm)^3$

2) $57.75(yd)^3$ 4) $350(in)^3$ 6) $1.2(in)^3$

Volume of Cylinder

1) $339.12(cm)^3$ 3) $602.88(yd)^3$ 5) $549.5(m)^3$

2) $549.5(mm)^3$ 4) $510.25(m)^3$ 6) $1,582.56(in)^3$

Volume of Pyramid and Cone

1) $326.67\ (in)^3$ 3) $1,050\ (in)^3$ 5) $7.15\ (in)^3$

2) $847.8\ (in)^3$ 4) $84.78(in)^3$ 6) $183.17(in)^3$

Surface Area Cubes

1) $600(in)^2$ 3) $337.5(in)^2$ 5) $37.5(in)^2$

2) $486(in)^2$ 4) $108(in)^2$ 6) $1,350(in)^2$

Surface Area Rectangle Prism

1) $262(in)^2$ 3) $256(in)^2$ 5) $135.5(in)^2$

2) $272.5(in)^2$ 4) $662(in)^2$ 6) $1,592(in)^2$

Surface Area Cylinder

1) $351.68(in)^2$ 3) $596.6(in)^2$ 5) $1,507.2(in)^2$

2) $408.2(in)^2$ 4) $62.8(in)^2$ 6) $449.02(in)^2$

Chapter 5 :
Exponent and Radicals

Positive Exponents

Simplify. Your answer should contain only positive exponents.

1) $4^4 =$

2) $3^5 =$

3) $\frac{5x^7 y}{xy} =$

4) $(11x^2 x)^3 =$

5) $(x^2)^6 =$

6) $\left(\frac{1}{5}\right)^3 =$

7) $0^{10} =$

8) $6 \times 6 \times 6 =$

9) $3 \times 3 \times 3 \times 3 \times 3 =$

10) $(4x^4 y)^2 =$

11) $9^3 =$

12) $(5x^3 y^2)^2 =$

13) $7 \times 10^4 =$

14) $0.3 \times 0.3 \times 0.3 =$

15) $\frac{1}{4} \times \frac{1}{4} \times \frac{1}{4} =$

16) $5^5 =$

17) $(5x^8 y^2)^3 =$

18) $8^3 =$

19) $y \times y \times y \times y =$

20) $8 \times 8 \times 8 \times 8 =$

21) $(2x^4 y^2 z)^3 =$

22) $10^0 =$

23) $(10x^4 y^{-1})^2 =$

24) $(3x^2 y^4)^3 =$

Negative Exponents

Simplify. Leave no negative exponents.

1) $2^{-3} =$

2) $7^{-2} =$

3) $\left(\frac{1}{4}\right)^{-2} =$

4) $5^{-3} =$

5) $1^{350} =$

6) $4^{-3} =$

7) $\left(\frac{1}{2}\right)^{-6} =$

8) $-8y^{-4} =$

9) $\left(\frac{1}{y^{-5}}\right)^{-3} =$

10) $x^{-\frac{4}{5}} =$

11) $\frac{1}{7^{-6}} =$

12) $3^{-5} =$

13) $5^{-2} =$

14) $13^{-2} =$

15) $30^{-2} =$

16) $x^{-8} =$

17) $(x^2)^{-4} =$

18) $x^{-2} \times x^{-2} \times x^{-2} \times x^{-2} =$

19) $\frac{1}{3} \times \frac{1}{3} =$

20) $100^{-2} =$

21) $100z^{-3} =$

22) $3^{-4} =$

23) $\left(-\frac{1}{13}\right)^{2} =$

24) $12^{0} =$

25) $\left(\frac{1}{x}\right)^{-12} =$

26) $10^{-2} =$

Add and subtract Exponents

Solve each problem.

1) $3^2 + 4^3 =$

2) $x^{10} + x^{10} =$

3) $5b^3 - 4b^3 =$

4) $6 + 5^2 =$

5) $9 - 6^2 =$

6) $12 + 3^2 =$

7) $5x^2 + 8x^2 =$

8) $9^2 + 2^6 =$

9) $3^6 - 4^3 =$

10) $8^2 - 10^0 =$

11) $7^2 - 4^2 =$

12) $9^2 + 3^4 =$

13) $12^2 - 5^2 =$

14) $7^2 + 7^2 =$

15) $7^3 - 5^3 =$

16) $1^{24} + 1^{28} =$

17) $4^3 - 2^3 =$

18) $5^4 - 5^2 =$

19) $7^2 - 4^2 =$

20) $5^2 + 8^2 =$

21) $4^2 + 3^4 =$

22) $18 + 2^4 =$

23) $7x^8 + 5x^8 =$

24) $9^0 + 8^2 =$

25) $8^2 + 8^2 =$

26) $6^3 + 3^2 =$

27) $(\frac{1}{2})^2 + (\frac{1}{2})^2 =$

28) $10^2 + 3^2 =$

Exponent multiplication

Simplify each of the following

1) $2^7 \times 2^4 =$

2) $5^3 \times 9^0 =$

3) $8^1 \times 3^2 =$

4) $a^{-7} \times a^{-7} =$

5) $y^{-3} \times y^{-3} \times y^{-3} =$

6) $4^5 \times 5^7 \times 4^{-4} \times 5^{-6} =$

7) $6x^4y^3 \times 4x^3y^2 =$

8) $(x^3)^5 =$

9) $(x^4y^6)^5 \times (x^4y^5)^{-5} =$

10) $8^4 \times 8^2 =$

11) $a^{4b} \times a^0 =$

12) $4^2 \times 4^2 =$

13) $a^{3m} \times a^{2n} =$

14) $2a^n \times 4b^n =$

15) $5^{-3} \times 4^{-3} =$

16) $6^{10} \times 3^{10} =$

17) $(7^6)^5 =$

18) $\left(\frac{1}{6}\right)^2 \times \left(\frac{1}{6}\right)^4 \times \left(\frac{1}{6}\right)^5 =$

19) $\left(\frac{1}{9}\right)^{52} \times 9^{52} =$

20) $(4m)^{\frac{4}{5}} \times (-2m)^{\frac{4}{5}} =$

21) $(x^4y)^{\frac{1}{4}} \times (xy^3)^{\frac{1}{4}} =$

22) $(2a^m b^n)^r =$

23) $(5x^3y^2)^3 =$

24) $(x^{\frac{1}{3}}y^2)^{\frac{-1}{3}} \times (x^4y^6)^0 =$

25) $7^8 \times 7^7 =$

26) $28^{\frac{1}{6}} \times 28^{\frac{1}{3}} =$

27) $9^4 \times 3^4 =$

28) $(x^{12})^0 =$

Exponent division

Simplify. Your answer should contain only positive exponents.

1) $\dfrac{7^6}{7} =$

2) $\dfrac{51x^4}{x} =$

3) $\dfrac{a^m}{a^{2n}} =$

4) $\dfrac{3x^{-6}}{15x^{-4}} =$

5) $\dfrac{63x^9}{7x^4} =$

6) $\dfrac{17x^7}{5x^8} =$

7) $\dfrac{36x^8}{12y^3} =$

8) $\dfrac{45xy^6}{x^4y^2} =$

9) $\dfrac{3x^9}{8x} =$

10) $\dfrac{45x^7y^9}{5x^8} =$

11) $\dfrac{12x^4}{20x^9y^{12}} =$

12) $\dfrac{8yx^7}{40yx^{10}} =$

13) $\dfrac{21x^3y^2}{3x^2y^3} =$

14) $\dfrac{x^{4.75}}{x^{0.75}} =$

15) $\dfrac{9x^4y}{18xy^3} =$

16) $\dfrac{34b^3r^8}{17a^2b^5} =$

17) $\dfrac{30x^7}{15x^9} =$

18) $\dfrac{44x^5}{11x^8} =$

19) $\dfrac{6^5}{6^3} =$

20) $\dfrac{x}{x^{15}} =$

21) $\dfrac{11^7}{11^4} =$

22) $\dfrac{5xy^5}{10y^3} =$

23) $\dfrac{11x^6y}{121xy^3} =$

24) $\dfrac{64x^5}{8y^9} =$

Scientific Notation

Write each number in scientific notation.

1) 8,500,000=

2) 700 =

3) 0.000008 =

4) 587,000 =

5) 0.00139 =

6) 0.85 =

7) 0.000093 =

8) 20,000,000 =

9) 28,000,000 =

10) 230,000,000 =

11) 0.000049 =

12) 0.00002 =

13) 0.00027 =

14) 70,000 =

15) 2,870 =

16) 190,000 =

17) 0.0223 =

18) 0.7 =

19) 0.082 =

20) 310,000 =

21) 48,000 =

22) 0.000098 =

23) 0.035 =

24) 1,778 =

25) 58,792 =

26) 24,600 =

27) 34,021 =

28) 9,100,000 =

Square Roots

Find the square root of each number.

1) $\sqrt{36} =$

2) $\sqrt{0} =$

3) $\sqrt{289} =$

4) $\sqrt{484} =$

5) $\sqrt{1,600} =$

6) $\sqrt{529} =$

7) $\sqrt{0.01} =$

8) $\sqrt{10,000} =$

9) $\sqrt{0.16} =$

10) $\sqrt{0.36} =$

11) $\sqrt{0.25} =$

12) $\sqrt{1.21} =$

13) $\sqrt{784} =$

14) $\sqrt{576} =$

15) $\sqrt{676} =$

16) $\sqrt{961} =$

17) $\sqrt{1,681} =$

18) $\sqrt{0.81} =$

19) $\sqrt{0.49} =$

20) $\sqrt{0.64} =$

21) $\sqrt{1,089} =$

22) $\sqrt{2,500} =$

23) $\sqrt{8,100} =$

24) $\sqrt{12,100} =$

25) $\sqrt{2.25} =$

26) $\sqrt{2.56} =$

27) $\sqrt{1.21} =$

28) $\sqrt{0.09} =$

Simplify Square Roots

Simplify the following.

1) $\sqrt{108} =$

2) $\sqrt{116} =$

3) $\sqrt{24} =$

4) $\sqrt{99} =$

5) $\sqrt{200} =$

6) $\sqrt{45} =$

7) $8\sqrt{50} =$

8) $3\sqrt{300} =$

9) $\sqrt{24} =$

10) $2\sqrt{18} =$

11) $4\sqrt{3} + 7\sqrt{3} =$

12) $\frac{11}{4+\sqrt{5}} =$

13) $\sqrt{48} =$

14) $\frac{4}{3-\sqrt{5}} =$

15) $\sqrt{18} \times \sqrt{2} =$

16) $\frac{\sqrt{300}}{\sqrt{3}} =$

17) $\frac{\sqrt{90}}{\sqrt{18 \times 5}} =$

18) $\sqrt{80y^6} =$

19) $6\sqrt{81a} =$

20) $\sqrt{41+8} + \sqrt{9} =$

21) $\sqrt{72} =$

22) $\sqrt{432} =$

23) $\sqrt{80} =$

24) $\sqrt{192} =$

25) $\sqrt{1,280} =$

26) $\sqrt{196} =$

Answers of Worksheets

Positive Exponents

1) 256

2) 243

3) $5x^6$

4) $1,331x^9$

5) x^{12}

6) $\frac{1}{125}$

7) 0

8) 6^3

9) 3^5

10) $16x^8y^2$

11) 729

12) $25x^6y^4$

13) 70,000

14) 0.3^3

15) $(\frac{1}{4})^3$

16) 3,125

17) $125x^{24}y^6$

18) 512

19) y^4

20) 8^4

21) $8x^{12}y^6z^3$

22) 1

23) $\frac{100x^8}{y^2}$

24) $27x^6y^{12}$

Negative Exponents

1) $\frac{1}{8}$

2) $\frac{1}{49}$

3) 16

4) $\frac{1}{125}$

5) 1

6) $\frac{1}{64}$

7) 64

8) $\frac{-8}{y^4}$

9) y^{15}

10) $\frac{1}{x^{\frac{4}{5}}}$

11) 7^6

12) $\frac{1}{243}$

13) $\frac{1}{25}$

14) $\frac{1}{169}$

15) $\frac{1}{900}$

16) $\frac{1}{x^8}$

17) $\frac{1}{x^8}$

18) $\frac{1}{x^8}$

19) $\frac{1}{3^2}$

20) $\frac{1}{10,000}$

21) $\frac{100}{z^3}$

22) $\frac{1}{81}$

23) $\frac{1}{169}$

24) 1

25) x^{12}

26) $\frac{1}{100}$

Add and subtract Exponents

1) 73

2) $2x^{10}$

3) b^3

4) 31

5) -27

6) 21

7) $13x^2$

8) 145

9) 665

10) 63

11) 33

12) 162

13) 119

14) 98

15) 218

16) 1

17) 56

18) 600

19) 33

20) 89

21) 97

22) 34

23) $12x^8$

24) 65

25) 128

26) 225

27) $\frac{1}{2}$

28) 109

Exponent multiplication

1) 2^{11}

2) 125

3) 72

4) a^{-14}

5) y^{-9}

6) 20

7) $24x^7y^5$

8) x^{15}

9) y^5

10) 8^6

11) a^{4b}

12) 4^4

13) a^{3m+2n}

14) $8(ab)^n$

15) 20^{-3}

16) 18^{10}

17) 7^{30}

18) $(\frac{1}{6})^{11}$

19) 1

20) $(-8m^2)^{\frac{4}{5}}$

21) $x^{\frac{5}{4}}y$

22) $2^r a^{mr} b^{nr}$

23) $125x^9y^6$

24) $x^{\frac{5}{4}}y$

25) 7^{15}

26) $28^{\frac{1}{2}}$

27) $27^4 = 3^{12}$

28) 1

Exponent division

1) 7^5

2) $51x^3$

3) a^{m-2n}

4) $\frac{1}{5x^2}$

5) $9x^5$

6) $\frac{17}{5x}$

7) $\frac{3x^8}{y^3}$

8) $\frac{45y^4}{x^3}$

9) $\frac{3x^8}{8}$

10) $\frac{9y^9}{x}$

11) $\frac{3}{5x^5y^{12}}$

12) $\frac{1}{5x^3}$

13) $\frac{7x}{y}$

14) x^4

15) $\frac{x^3}{2y^2}$

16) $\frac{2r^8}{a^2b^2}$

17) $\frac{2}{x^2}$

18) $\frac{4}{x^3}$

19) 6^2

20) $\frac{1}{x^{14}}$

21) 11^3

22) $\frac{1}{2}xy^2$

23) $\frac{x^5}{11y^2}$

24) $\frac{8x^5}{y^9}$

Scientific Notation

1) 8.5×10^6

2) 7×10^2

3) 8×10^{-6}

4) 5.87×10^5

5) 1.39×10^{-3}

6) 8.5×10^{-1}

7) 9.3×10^{-5}

8) 2×10^7

9) 2.8×10^7

10) 2.3×10^8

11) 4.9×10^{-5}

12) 2×10^{-5}

13) 2.7×10^{-4}

14) 7×10^4

15) 2.87×10^3

16) 1.9×10^5

17) 2.23×10^{-2}

18) 7×10^{-1}

19) 8.2×10^{-2}

20) 3.1×10^5

21) 4.8×10^4

22) 9.8×10^{-5}

23) 3.5×10^{-2}

24) 1.778×10^3

25) 5.8792×10^4

26) 2.46×10^4

27) 3.4021×10^4

28) 9.1×10^6

Square Roots

1) 6

2) 0

3) 17

4) 22

5) 40

6) 23

7) 0.1

8) 100

9) 0.4

10) 0.6

11) 0.5

12) 1.1

13) 28

14) 24

15) 26

16) 31

17) 41

18) 0.9

19) 0.7

20) 0.8

21) 33

22) 50

23) 90

24) 110

25) 1.5

26) 1.6

27) 1.1

28) 0.3

Simplify Square Roots

1) $6\sqrt{3}$

2) $2\sqrt{29}$

3) $2\sqrt{6}$

4) $3\sqrt{11}$

5) $10\sqrt{2}$

6) $3\sqrt{5}$

7) $40\sqrt{2}$

8) $30\sqrt{3}$

9) $2\sqrt{6}$

10) $6\sqrt{2}$

11) $11\sqrt{3}$

12) $4 - \sqrt{5}$

13) $4\sqrt{3}$

14) $3 + \sqrt{5}$

15) 6

16) 10

17) 1

18) $4y^3\sqrt{5}$

19) $54\sqrt{a}$

20) 10

21) $6\sqrt{2}$

22) $12\sqrt{3}$

23) $4\sqrt{5}$

24) $8\sqrt{3}$

25) $16\sqrt{5}$

26) 14

Chapter 6 :
Ratio, Proportion and Percent

Proportions

Find a missing number in a proportion.

1) $\dfrac{3}{5} = \dfrac{18}{a}$

11) $\dfrac{10}{8} = \dfrac{5}{a}$

2) $\dfrac{a}{8} = \dfrac{25}{40}$

12) $\dfrac{15}{a} = \dfrac{3}{13}$

3) $\dfrac{24}{120} = \dfrac{a}{10}$

13) $\dfrac{4}{11} = \dfrac{a}{12}$

4) $\dfrac{16}{a} = \dfrac{96}{36}$

14) $\dfrac{\sqrt{36}}{3} = \dfrac{48}{a}$

5) $\dfrac{4}{a} = \dfrac{16}{75}$

15) $\dfrac{6}{a} = \dfrac{6.6}{39.6}$

6) $\dfrac{\sqrt{9}}{4} = \dfrac{a}{32}$

16) $\dfrac{60}{140} = \dfrac{a}{280}$

7) $\dfrac{2}{4} = \dfrac{18}{a}$

17) $\dfrac{42}{200} = \dfrac{a}{68}$

8) $\dfrac{7}{14} = \dfrac{a}{35}$

18) $\dfrac{23}{161} = \dfrac{a}{7}$

9) $\dfrac{7}{a} = \dfrac{4.2}{6}$

19) $\dfrac{10}{32} = \dfrac{4}{a}$

10) $\dfrac{2}{12} = \dfrac{8}{a}$

20) $\dfrac{18}{14} = \dfrac{27}{a}$

Reduce Ratio

Reduce each ratio to the simplest form.

1) 6 : 24 =

2) 7 : 42 =

3) 72 : 40 =

4) 30 : 25 =

5) 12 : 120 =

6) 16 : 2 =

7) 70 : 350 =

8) 4 : 144 =

9) 25 : 75 =

10) 4.8 : 5.6 =

11) 110 : 330 =

12) 3 : 5 =

13) 60 : 100 =

14) 24 : 36 =

15) 34 : 68 =

16) 32 : 8 =

17) 140 : 35 =

18) 20 : 200 =

19) 126 : 84 =

20) 156 : 198 =

21) 30 : 60 =

22) 12 : 14 =

23) 10 : 150 =

24) 15 : 90 =

Percent

Find the Percent of Numbers.

1) 30% of 45 =

2) 23% of 16 =

3) 15% of 17 =

4) 12% of 140 =

5) 7% of 70 =

6) 35% of 12 =

7) 18% of 5 =

8) 12% of 46 =

9) 40% of 62 =

10) 4.5% of 50 =

11) 85% of 18 =

12) 60% of 50 =

13) 18% of 180 =

14) 2% of 240 =

15) 75% of 0 =

16) 80% of 120 =

17) 36% of 45 =

18) 10% of 70 =

19) 8% of 13 =

20) 4% of 8 =

21) 30% of 44 =

22) 80% of 17 =

23) 22% of 35 =

24) 8% of 150 =

25) 40% of 270 =

26) 6% of 15 =

27) 9% of 360 =

28) 10% of 56 =

Discount, Tax and Tip

Find the selling price of each item.

1) Original price of a computer: $300

Tax: 7%, Selling price: $_____

2) Original price of a laptop: $210

Tax: 16%, Selling price: $_____

3) Original price of a sofa: $400

Tax: 8%, Selling price: $_____

4) Original price of a car: $12,600

Tax: 3.5%, Selling price: $_____

5) Original price of a Table: $500

Tax: 12%, Selling price: $_____

6) Original price of a house: $280,000

Tax: 1.5%, Selling price: $_____

7) Original price of a tablet: $460

Discount: 30%, Selling price: $____

8) Original price of a chair: $110

Discount: 10%, Selling price: $____

9) Original price of a book: $30

Discount: 10% Selling price: $____

10) Original price of a cellphone: 720

Discount: 12% Selling price: $_____

11) Food bill: $42

Tip: 10%   Price: $_____

12) Food bill: $38

Tipp: 15%   Price: $_____

13) Food bill: $80

Tip: 27%   Price: $_____

14) Food bill: $62

Tipp: 15%   Price: $_____

Find the answer for each word problem.

15) Nicolas hired a moving company. The company charged $600 for its services, and Nicolas gives the movers a 15% tip. How much does Nicolas tip the movers? $_____

16) Mason has lunch at a restaurant and the cost of his meal is $70. Mason wants to leave a 7% tip. What is Mason's total bill including tip? $_____

Percent of Change

Find each percent of change.

1) From 400 to 800. ___%

2) From 90 ft to 360 ft. ___%

3) From $120 to $840. ___%

4) From 50 cm to 150 cm. ___%

5) From 10 to 30. ___%

6) From 60 to 108. ___%

7) From 120 to 180. ___%

8) From 400 to 600. ___%

9) From 170 to 238. ___%

10) From 200 to 350. ___%

Calculate each percent of change word problem.

11) Bob got a raise, and his hourly wage increased from $32 to $40. What is the percent increase? ____%

12) The price of a pair of shoes increases from $50 to $80. What is the percent increase? ___%

13) At a coffee shop, the price of a cup of coffee increased from $3.50 to $4.2. What is the percent increase in the cost of the coffee? _____%

14) 22cm are cut from a 88 cm board. What is the percent decrease in length? _%

15) In a class, the number of students has been increased from 112 to 168. What is the percent increase? _____%

16) The price of gasoline rose from $36.8 to $42.32 in one month. By what percent did the gas price rise? _____%

17) A shirt was originally priced at $42. It went on sale for $33.6. What was the percent that the shirt was discounted? _____%

Simple Interest

Determine the simple interest for these loans.

1) $180 at 15% for 5 years. $ _____

2) $1,600 at 4% for 2 years. $ _____

3) $900 at 25% for 4 years. $ _____

4) $9,200 at 1.5% for 8 months. $ ___

5) $600 at 5% for 7 months. $ _____

6) $40,000 at 8.5% for 3 years. $ ____

7) $7,400 at 8% for 5 years. $ _____

8) $500 at 6.5% for 2 years. $ _____

9) $800 at 4.5 % for 4 months. $ ____

10) $6,000 at 3.5% for 5 years. $ ___

Calculate each simple interest word problem.

11) A new car, valued at $18,000, depreciates at 5.5% per year. What is the value of the car one year after purchase? $_____

12) Sara puts $9,000 into an investment yielding 8% annual simple interest; she left the money in for two years. How much interest does Sara get at the end of those three years? $_____

13) A bank is offering 12.5% simple interest on a savings account. If you deposit $30,400, how much interest will you earn in four years? $_____

14) $1,200 interest is earned on a principal of $5,000 at a simple interest rate of 12% interest per year. For how many years was the principal invested? _____

15) In how many years will $1,800 yield an interest of $432 at 6% simple interest? _____

16) Jim invested $8,000 in a bond at a yearly rate of 2.5%. He earned $600 in interest. How long was the money invested? _____

Answers of Worksheets

Proportions

1) $a = 30$
2) $a = 5$
3) $a = 2$
4) $a = 6$
5) $a = 18.75$
6) $a = 24$
7) $a = 36$
8) $a = 17.5$
9) $a = 10$
10) $a = 48$
11) $a = 4$
12) $a = 65$
13) $a = \frac{48}{11}$
14) $a = 24$
15) $a = 36$
16) $a = 120$
17) $a = 14.28$
18) $a = 1$
19) $a = 12.8$
20) $a = 21$

Reduce Ratio

1) $1:4$
2) $1:6$
3) $9:5$
4) $6:5$
5) $1:10$
6) $8:1$
7) $1:5$
8) $1:36$
9) $1:3$
10) $0.6:0.7$
11) $11:33$
12) $0.6:1$
13) $3:5$
14) $2:3$
15) $1:2$
16) $4:1$
17) $4:1$
18) $1:10$
19) $3:2$
20) $26:33$
21) $1:2$
22) $6:7$
23) $1:15$
24) $1:6$

Percent

2) 13.5
3) 3.68
4) 2.55
5) 16.8
6) 4.9
7) 4.2
8) 0.9
9) 5.52
10) 24.8
11) 2.25
12) 15.3
13) 30
14) 13.4
15) 4.8
16) 0
17) 96
18) 16.2
19) 7
20) 1.04
21) 0.32
22) 13.2
23) 13.6
24) 7.7
25) 12
26) 108
27) 0.9
28) 32.4
29) 5.6

Discount, Tax and Tip

1) $321.00
2) $243.6
3) $432.00
4) $13,041.00
5) $560.00
6) $284,200

7) $322.00
8) $99.00
9) $27.00
10) $633.60
11) $46.2
12) $43.7

13) $101.60
14) $71.3
15) $90.00
16) $74.90

Percent of Change

1) 100%
2) 400%
3) 600%
4) 300%
5) 200%
6) 80%

7) 50%
8) 50%
9) 40%
10) 75%
11) 25%
12) 60%

13) 20%
14) 25%
15) 50%
16) 15%
17) 20%

Simple Interest

1) $135.00
2) $128.00
3) $900.00
4) $92.00
5) $17.50
6) $10,200.00

7) $2,960.00
8) $65.00
9) $144.00
10) $1,050.00
11) $17,010.00
12) $2,160.00

13) $15,200.00
14) 2 years
15) 4 years
16) 3 years

Chapter 7 :

Transformations

Translations

Graph the image of the figure using the transformation given.

1) Translation: 5 units right and 2 units down

2) Translation: 1 units left and 3 units down

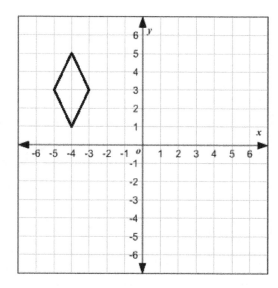

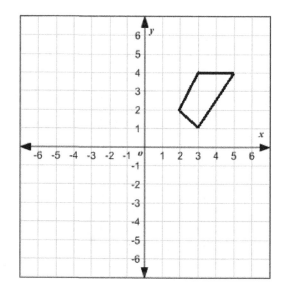

Write a rule to describe each transformation.

3)

4)

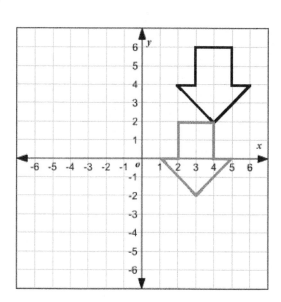

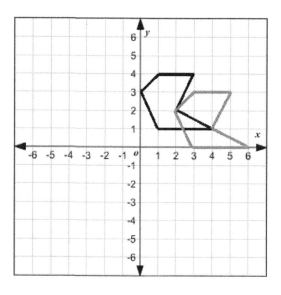

Reflections

Graph the image of the figure using the transformation given.

1) Reflection across $x = 3$

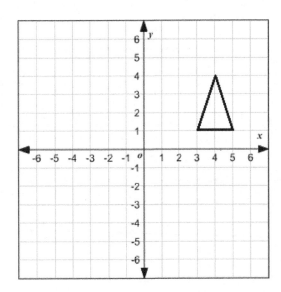

2) Reflection across $y = x$

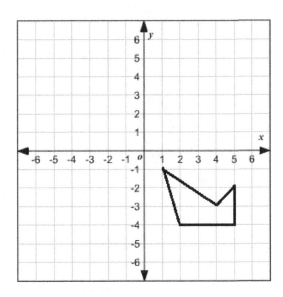

3) Reflection across $y = -2$

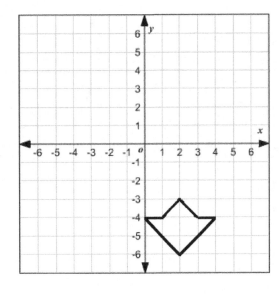

4) Reflection across y axis

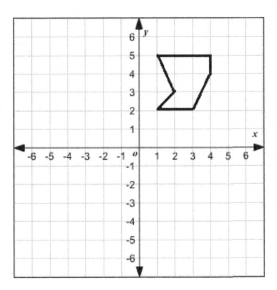

Write a rule to describe each transformation.

5)

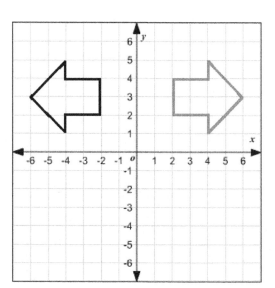

6)

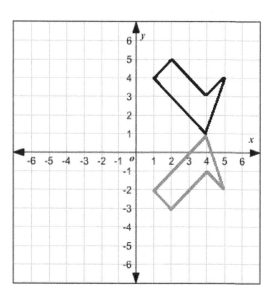

7)

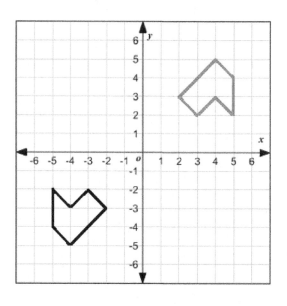

8)

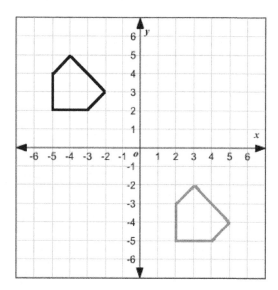

Rotations

Graph the image of the figure using the transformation given.

1) Rotation 90° counterclockwise about the origin

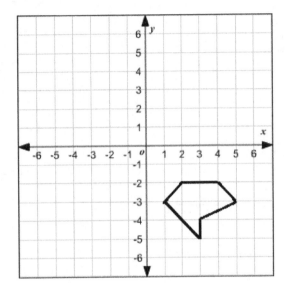

2) Rotation 270° counterclockwise about the origin

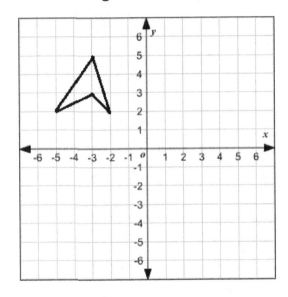

3) Rotation 270° clockwise about the origin

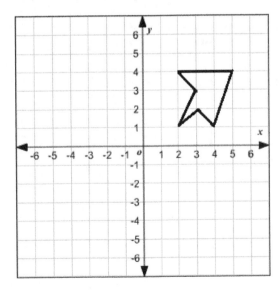

4) Rotation 180° counterclockwise about the origin

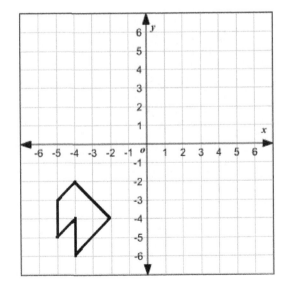

Write a rule to describe each transformation.

5)

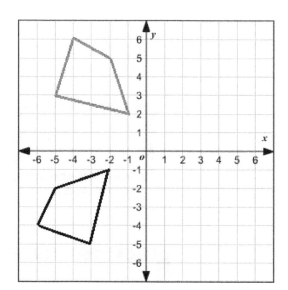

6)

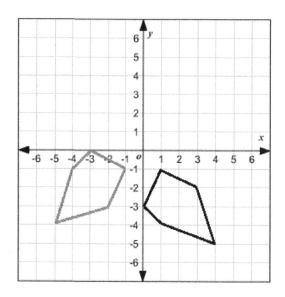

7)

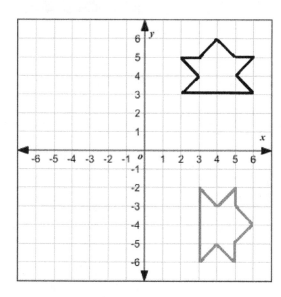

8)

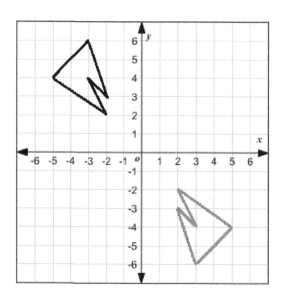

Dilations

Determine whether the dilation from figure M to figure N is a reduction or an enlargement. Then find the scale factor and the missing length.

1)

12 cm

x cm   N

18 cm

M

6 cm

2)

5 cm

x cm   M

15cm

9 cm   N

✍ *Draw a dilation of the figure using the given scale factor.*

3) $k = \frac{1}{2}$

4) $k = 3$

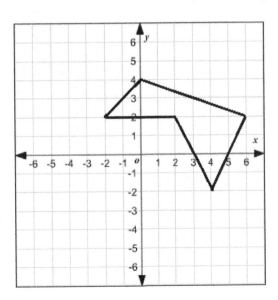

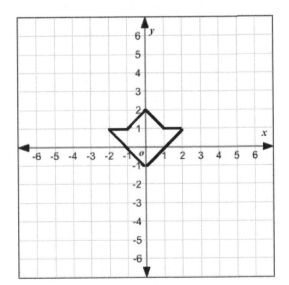

Coordinates of Vertices

Calculate the new coordinates after the given transformations.

1) Translate: 1-unit right and 2 units down.

   $A(-1, 0), B(1, -3), C(2, 1)$

2) Rotation: 270° clockwise about the origin.

   $D(-1, 1), E(-5, 6), F(-7, 3), G(-3, 10)$

3) Rotation: 180° counterclockwise about the origin.

   $P(3, 0), Q(5, 2), R(-1, 4), S(2, -7)$

4) Rotation: 90° clockwise about the origin.

   $J(-2, 5), K(-5, 0), L(3, -6)$

5) Reflection: over the y axis.

   $C(-2, -6), D(4, -3), W(1, -7), Y(5, 1)$

6) Reflection: across the line $y = -x$.

   $A(3, -2), B(8, -4), C(6, -6), D(1, -5)$

7) Reflection: across the line $y = -2$.

   $K(-1, 1), L(-4, 2), M(4, -1), N(2, 3)$

8) Dilate: Reduction by scale factor $\frac{1}{3}$.

   $A(6, 3), B(-12, 0), C(-9, 6)$

9) Dilate: Enlargement by scale factor 2.

   $F(-1, 4), G(-3, 0), H(3, 2)$

Answers of Worksheets

Translations

1)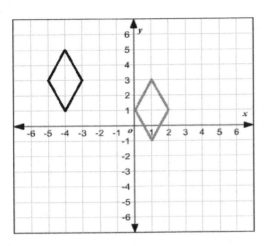

2)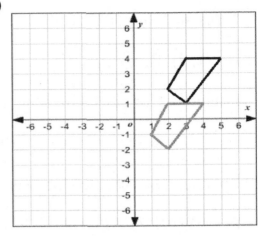

3) Translation: : 1 units left and 4 units down

4) Translation: 2 units right and 1 unit down

Reflections

1)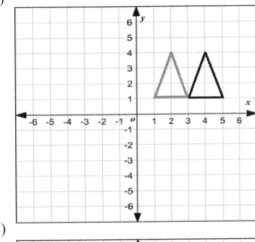

2)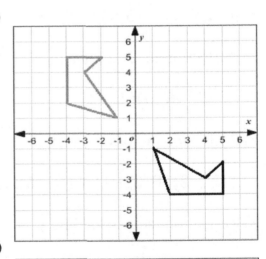

3)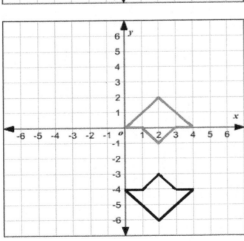

4)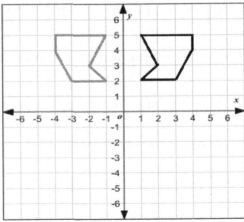

5) Reflection across the x = 0 (y axis)

6) Reflection across the y = 1

7) Reflection against the origin

8) Reflection across the y = x

Rotations

1)

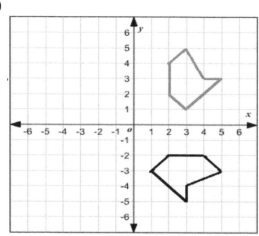

2)

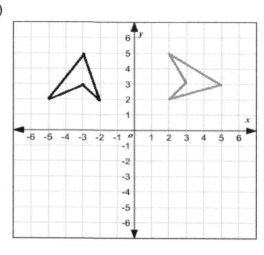

3)

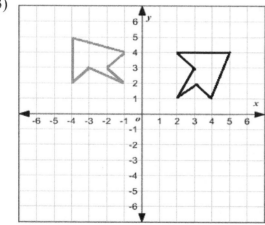

4)

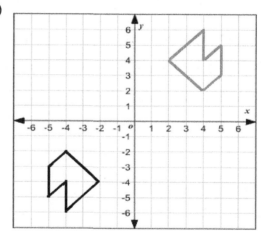

5) Rotation 90° clockwise about the origin

6) Rotation 180° about the origin

7) Rotation 270° counter clockwise about the origin

8) Rotation 90° clockwise about the origin

Dilations

1) Reduction, $k = \dfrac{3}{2}$, $x = 4\ cm$

2) Enlargement, $k = \dfrac{1}{3}$, $x = 3\ cm$

3)

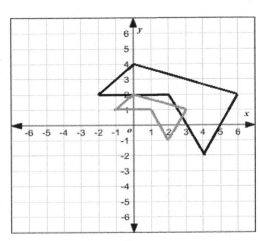

4)

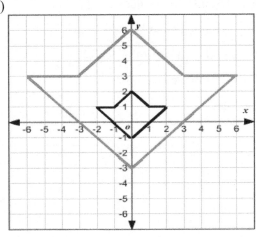

Coordinate of Vertices

1) $A'(0, -2), B'(2, -5), C'(3, -1)$

2) $D'(-1, -1), E'(-6, -5), F'(-3, -7), G'(-10, -3)$

3) $P'(-3, 0), Q'(-5, -2), R'(1, -4), S'(-2, 7)$

4) $J'(5, 2), K'(0, 5), L'(-6, -3)$

5) $C'(2, -6), D'(-4, -3), W'(-1, -7), Y'(-5, 1)$

6) $A'(2, -3), B'(4, -8), C'(6, -6), D'(5, -1)$

7) $K'(-1, -5), L'(-4, -6), M'(4, -3), N'(2, -7)$

8) $A'(2, 1), B'(-4, 0), C'(-3, 2)$

9) $F'(-2, 8), G'(-6, 0), H'(6, 4)$

Chapter 8 :

Equations and

Inequality

Distributive and Simplifying Expressions

Simplify each expression.

1) $4x + 4 - 9 =$

2) $-(-5 - 7x) =$

3) $(-2x + 5)(-3) =$

4) $(-3x)(x + 4) =$

5) $-2x + x^2 + 4x^2 =$

6) $7y + 7x + 8y - 5x =$

7) $-3x + 3y + 14x - 9y =$

8) $-2x - 5 + 8x + \frac{16}{4} =$

9) $5 - 8(x - 2) =$

10) $-5 - 5x + 3x =$

11) $(x - 3y)2 + 4y =$

12) $2.5x^2 \times (-5x) =$

13) $-4 - 2x^2 + 6x^2 =$

14) $8 + 14x^2 + 4 =$

15) $2(-2x - 5) + 12 =$

16) $(-x)(-2 + 4x) - x(6 + x) =$

17) $-3(6 + 12) - 3x + 5x =$

18) $-4(5 - 12x - 3x) =$

19) $3(-2x - 6) =$

20) $21 + 8x - 21 =$

21) $x(-4x + 7) =$

22) $5xy + 4x - 3y + x + 2y =$

23) $3(-x - 7) + 9 =$

24) $(-3x - 8) + 12 =$

25) $3x + 4y - 6 + 1 =$

26) $(-2 + 4x) - 4x(1 + 3x) =$

27) $(-4)(-2x - 2y) =$

28) $6(-x - 3) + 4 =$

Factoring Expressions

Factor the common factor out of each expression.

1) $15x - 12 =$

2) $3x - 12 =$

3) $\frac{45}{15}x - 24 =$

4) $6b - 30 =$

5) $4a^2 - 24a =$

6) $2xy - 10y =$

7) $5x^2y + 15x =$

8) $a^2 - 8a + 7ab =$

9) $2a^2 + 2ab =$

10) $4x + 20 =$

11) $24x - 36xy =$

12) $8x - 6 =$

13) $\frac{1}{4}x - \frac{3}{4}y =$

14) $7xy - \frac{14}{3}x =$

15) $4ab + 12c =$

16) $\frac{1}{5}x - \frac{4}{5} =$

17) $12x - 18xy =$

18) $x^2 + 8x =$

19) $4x^2 - 12y =$

20) $4x^3 + 3xy + x^2 =$

21) $21x - 14 =$

22) $20b - 60c + 20d =$

23) $24ab - 8ac =$

24) $ax - ay - 3x + 3y =$

25) $3ax + 4a + 9x + 12 =$

26) $x^2 - 15x =$

27) $7x^3 - 14x^2 =$

28) $5x^2 - 60xy =$

Evaluate One Variable Expressions

Evaluate each using the values given.

1) $x + 5x, x = 2$

2) $5(-7 + 4x), x = 1$

3) $4x + 6x, x = -1$

4) $4(2 - x) + 4, x = 2$

5) $8x + 2x - 12, x = 2$

6) $5x + 11x + 12, x = -1$

7) $5x - 2x - 4, x = 5$

8) $\frac{3(5x+8)}{9}, x = 2$

9) $2x - 85, x = 32$

10) $\frac{x}{18}, \; x = 108$

11) $7(3 + 2x) - 33, x = 5$

12) $7(x + 3) - 23, x = 4$

13) $\frac{x+(-6)}{-3}, x = -6$

14) $8(6 - 3x) + 5, x = 2$

15) $-5 - \frac{2x}{10} + 6x, x = 10$

16) $5x + 11x, x = 1$

17) $-12x + 3(5 + 3x), x = -7$

18) $x + 11x, x = 0.5$

19) $\frac{(2x-2)}{6}, x = 13$

20) $3(-1 - 2x), x = 5$

21) $5x - (5 - x), x = 3$

22) $\left(-\frac{15}{x}\right) + 2 + x, x = 5$

23) $-\frac{x \times 7}{x}, x = 8$

24) $2(-1 - 3x), x = 2$

25) $3x^2 + 8x, x = 1$

26) $2(5x + 2) - 3(x - 3), x = 2$

27) $-5x - 5, x = -5$

28) $6x + 3x, x = 2$

Evaluate Two Variable Expressions

Evaluate the expressions.

1) $x + 6y$, $x = 3, y = 5$

2) $(-2)(-2x - 3y)$, $x = 1, y = 1$

3) $3x + 4y$, $x = 6, y = 2$

4) $\frac{x-4}{y+1}$, $x = 18, y = 1$

5) $\frac{a}{8} - 6b$, $a = 32, b = 4$

6) $3x - 4(y - 8)$, $x = 5, y = 3$

7) $3x + 2y - 10$, $x = 2, y = 10$

8) $-3x + 10 + 8y - 5$, $x = 2, y = 1$

9) $yx \div 3$, $x = 9, y = 9$

10) $a - b \div 3$, $a = 3, b = 12$

11) $6(x - y)$, $x = 7, y = 4$

12) $5x - 4y$, $x = 5, y = 8$

13) $\frac{10}{a} + 3b$, $a = 5, b = 4$

14) $2x^2 + 4xy$, $x = 3, y = 5$

15) $10 - \frac{xy}{12} + y$, $x = 4, y = 3$

16) $5(3x - y)$, $x = 4, y = -6$

17) $5x^2 - 3y^2$, $x = -1, y = 2$

18) $5x + \frac{y}{4}$, $x = 6, y = 16$

19) $4(4x - 2y)$, $x = 3, y = 5$

20) $4x(y - \frac{1}{2})$, $x = 5, y = 4$

21) $5(x^2 - 2y)$, $x = 3, y = 2$

22) $5xy$, $x = 2, y = 8$

23) $\frac{1}{3}y^3\left(y - \frac{1}{4}x\right)$, $x = -4, y = 3$

24) $-3(x - 5y) - 2x$, $x = 4, y = 2$

25) $-2x + \frac{1}{6}xy$, $x = 3, y = 6$

26) $x^2 + xy^2$, $x = 6, y = 5$

27) $x - 3y + 6$, $x = 8, y = 5$

28) $\frac{xy}{4x+y}$, $x = 6, y = 5$

Graphing Linear Equation

Sketch the graph of each line.

1) $y = 2x - 5$ 2) $y = -2x + 3$ 3) $x - y = 0$

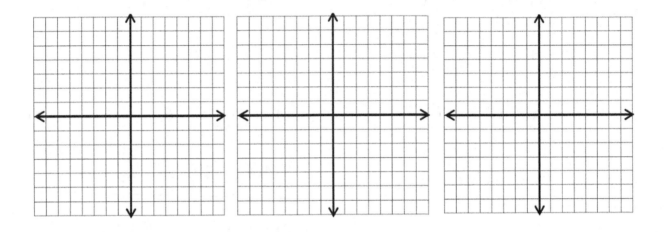

4) $x + y = 3$ 5) $5x + 3y = -2$ 6) $y - 3x + 2 = 0$

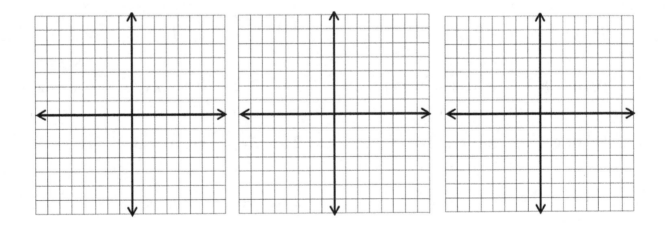

One Step Equations

Solve each equation.

1) $88 = (-24) + x$

2) $6x = (-81)$

3) $(-63) = (-7x)$

4) $(-8) = 3 + x$

5) $4 + \dfrac{x}{2} = (-3)$

6) $8x = (-104)$

7) $62 = x - 13$

8) $\dfrac{x}{3} = (-15)$

9) $x + 112 = 154$

10) $x - \dfrac{1}{3} = \dfrac{2}{3}$

11) $(-24) = x - 32$

12) $(-3x) = 39$

13) $(-169) = (13x)$

14) $-4x + 42 = 50$

15) $5x + 3 = 38$

16) $80 = (-8x)$

17) $3x + 7 = 19$

18) $11x = 121$

19) $x - 18 = 15$

20) $0.9x = 4.5$

21) $4x = 84$

22) $2x + 2.98 = 66.98$

23) $x + 9 = 6$

24) $x + 24 = 16$

25) $9x + 51 = 15$

26) $\dfrac{1}{6}x + 60 = 48$

Two Steps Equations

Solve each equation.

1) $12(3 + x) = 84$

2) $(-14)(x - 2) = 112$

3) $(-4)(3x - 4) = (-8)$

4) $15(2 + x) = -45$

5) $38(3x + 11) = 76$

6) $4(2x + 2) = 24$

7) $5(8 + 3x) = (-20)$

8) $(-5)(5x - 3) = 40$

9) $2x + 12 = 16$

10) $\frac{4x - 5}{5} = 3$

11) $(-3) = \frac{x + 4}{7}$

12) $80 = (-8)(x - 3)$

13) $\frac{x}{3} + 27 = 39$

14) $\frac{1}{8} = \frac{1}{4} + \frac{x}{8}$

15) $\frac{33 + 3x}{15} = (-6)$

16) $(-3)(10 + 5x) = (-15)$

17) $(-3x) + 12 = 24$

18) $\frac{x + 5}{5} = -5$

19) $\frac{x + 23}{8} = 3$

20) $(-4) + \frac{x}{2} = (-14)$

21) $-5 = \frac{x + 8}{6}$

22) $\frac{27x - 9}{18} = 4$

23) $\frac{2x - 12}{4} = 3$

24) $45 = (-5)(x - 45)$

Multi Steps Equations

Solve each equation.

1) $5 - (4 - 5x) = 6$

2) $-25 = -(4x + 17)$

3) $6x - 22 = (-2x) + 10$

4) $-75 = (-5x) - 10x$

5) $3(2 + 3x) + 3x = -30$

6) $5x - 18 = 2 + 2x - 7 + 2x$

7) $12 - 6x = (-36) - 3x + 3x$

8) $16 - 4x - 4x = 8 - 4x$

9) $8 + 7x + x = (-12) + 3x$

10) $(-3x) - 3(-2 + 4x) = 366$

11) $20 = (-200x) - 5 + 5$

12) $61 = 5x - 23 + 7x$

13) $14(4 + 2x) = 280$

14) $-60 = (-7x) - 13x$

15) $2(4x + 5) = -2(x + 4) - 22$

16) $11x - 17 = 6x + 8$

17) $9 = -3(x - 8)$

18) $(-6) - 8x = 6(1 + 2x)$

19) $x + 3 = -2(9 + 3x)$

20) $15 = 6 - 5x - 11$

21) $-15 - 9x - 3x = 12 - 3x$

22) $-30 - 3x + 5x = 20 - 23x$

23) $35 - 6x - 16x = -10 - 16x$

24) $15x - 17 = 6x + 10$

Graphing Linear Inequalities

Sketch the graph of each linear inequality.

1) $y > 2x - 3$ 2) $y < x + 3$ 3) $y \leq -3x - 8$

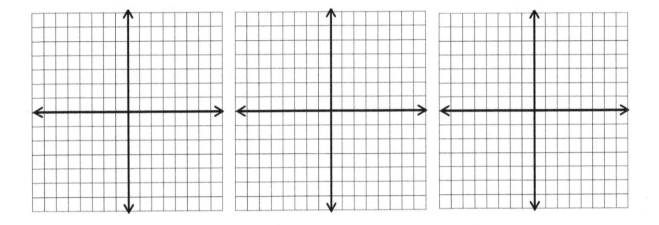

4) $3y \geq 6 + 3x$ 5) $-3y < x - 12$ 6) $2y \geq -8x + 4$

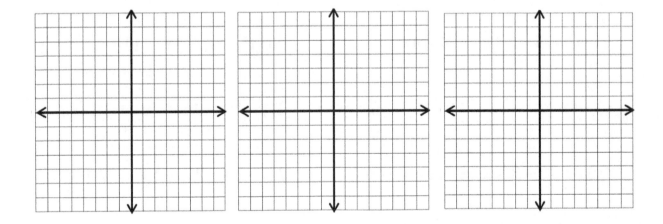

One Step Inequality

Solve each inequality.

1) $14x < 28$

2) $x + 17 \geq -4$

3) $x - 2 \leq 10$

4) $-2x + 4 > -10$

5) $x + 18 \geq -6$

6) $x + 9 \geq 5$

7) $x - \frac{1}{3} \leq 5$

8) $-7x < 42$

9) $-x + 8 > -3$

10) $\frac{x}{3} + 3 > -9$

11) $-x + 8 > -4$

12) $x - 14 \leq 18$

13) $-x - 7 \leq -8$

14) $x + 28 \geq -11$

15) $x + \frac{1}{3} \geq -\frac{2}{3}$

16) $x + 6 \geq -14$

17) $x - 42 \leq -48$

18) $x - 5 \leq 4$

19) $-x + 5 > -6$

20) $x + 6 \geq -12$

21) $8x + 6 \leq 22$

22) $6x - 3 \geq 9$

23) $9x - 5 < 22$

24) $6x - 7 \leq 35$

Two Steps Inequality

Solve each inequality

1) $4x - 6 \leq 14$

2) $9x - 12 \leq 24$

3) $\frac{-1}{4}x + \frac{x}{2} \leq \frac{1}{8}$

4) $10x + 20 \geq 60$

5) $8x - 14 \geq 18$

6) $3x - 5 \leq 16$

7) $8x - 2 \leq 14$

8) $9x + 5 \leq 23$

9) $2x + 10 > 32$

10) $\frac{x}{8} + 2 \leq 4$

11) $3x + 4 \geq 37$

12) $3x - 8 < 10$

13) $6 \geq \frac{x+7}{2}$

14) $3x + 9 < 48$

15) $\frac{4+x}{5} \geq 3$

16) $16 + 4x < 36$

17) $16 > 6x - 8$

18) $11 + \frac{x}{2} < 7$

19) $-8 + 8x > 48$

20) $6 + \frac{x}{9} < 3$

Multi Steps Inequality

Solve the inequalities.

1) $8x - 12 < 10x - 18$

11) $7x - 4 \leq 8x + 9$

2) $\dfrac{4x + 10}{6} \leq x$

12) $\dfrac{2x - 7}{5} > 2$

3) $14x - 10 > 6x + 30$

13) $8(x + 2) < 6x + 10$

4) $-3x > -6x + 4$

14) $-8x + 12 \leq 4(x - 9)$

5) $3 + \dfrac{x}{2} < \dfrac{x}{4}$

15) $\dfrac{5x - 6}{3} > 3x + 2$

6) $\dfrac{4x - 6}{8} > x$

16) $2(x - 8) + 10 \geq 4x - 2$

7) $4x - 20 + 4 > 6x - 8$

17) $\dfrac{-5x + 7}{6} > 5x$

8) $x - 8 > 11 + 3(x + 5)$

18) $-6x - 8 > -14x$

9) $\dfrac{x}{3} + 2 > x$

19) $\dfrac{1}{4}x - 16 > \dfrac{1}{8}x - 23$

10) $-7x + 8 \geq -6(4x - 8) - 8x$

20) $-16(x - 9) \leq 20x$

Systems of Equations

Calculate each system of equations.

1) $-12x + 14y = 16$ $x = $ ____
   $2x + 8y = 18$ $y = $ ____

2) $-8x + 24y = 24$ $x = $ ____
   $28x - 32y = 20$ $y = $ ____

3) $y = -9$ $x = $ ____
   $4x - 10y = 24$ $y = $ ____

4) $8y = -8x + 40$ $x = $ ____
   16
   $x - 4y = -24$ $y = $ ____

5) $10x - 9y = -13$ $x = $ ____
   $-5x + 3y = 11$ $y = $ ____

6) $-6x - 8y = 10$ $x = $ ____
   $4x - 8y = 20$ $y = $ ____

7) $5x - 14y = -23$ $x = $ ____
   $-6x + 7y = 8$ $y = $ ____

8) $-4x + 3y = 3$ $x = $ ____
   $-x + 2y = 5$ $y = $ ____

9) $-4x + 5y = 15$ $x = $ ____
   $-3x + 4y = -10$ $y = $ ____

10) $-6x - 6y = -21$ $x = $ ____
    $-6x + 6y = -66$ $y = $ ____

11) $12x - 21y = 6$ $x = $ ____
    $-6x - 3y = -12$ $y = $ ____

12) $-8x - 8y = -28$ $x = $ ____
    $8x - 8y = 88$ $y = $ ____

13) $8x + 10y = 6$ $x = $ ____
    $6x - 2y = 12$ $y = $ ____

14) $6x - 4y = 4$ $x = $ ____
    $20x - 20y = 40$ $y = $ ____

15) $10x + 16y = 28$ $x = $ ____
    $-6x - 4y = -6$ $y = $ ____

16) $16x + 10y = 8$ $x = $ ____
    $-6x - 8y = 30$ $y = $ ____

Systems of Equations Word Problems

Find the answer for each word problem.

1) Tickets to a movie cost $6 for adults and $4 for students. A group of friends purchased 9 tickets for $50.00. How many adults ticket did they buy? ____

2) At a store, Eva bought two shirts and five hats for $77.00. Nicole bought three same shirts and four same hats for $84.00. What is the price of each shirt? _____

3) A farmhouse shelters 10 animals, some are pigs, and some are ducks. Altogether there are 36 legs. How many pigs are there? _____

4) A class of 85 students went on a field trip. They took 24 vehicles, some cars and some buses. If each car holds 3 students and each bus hold 16 students, how many buses did they take? _____

5) A theater is selling tickets for a performance. Mr. Smith purchased 8 senior tickets and 10 child tickets for $248 for his friends and family. Mr. Jackson purchased 4 senior tickets and 6 child tickets for $132. What is the price of a senior ticket? $_____

6) The difference of two numbers is 15. Their sum is 33. What is the bigger number? $_____

7) The sum of the digits of a certain two–digit number is 7. Reversing its digits increase the number by 9. What is the number? _____

8) The difference of two numbers is 11. Their sum is 25. What are the numbers? _____

9) The length of a rectangle is 5 meters greater than 2 times the width. The perimeter of rectangle is 28 meters. What is the length of the rectangle? _____

10) Jim has 23 nickels and dimes totaling $2.40. How many nickels does he have? _____

Finding Distance of Two Points

Find the distance between each pair of points.

1) $(4, 2), (-2, -6)$

2) $(-8, -4), (8, 8)$

3) $(-6, 0), (30, 48)$

4) $(-8, -2), (2, 22)$

5) $(3, -2), (-6, -14)$

6) $(-6, 0), (-2, 3)$

7) $(3, 2), (11, 17)$

8) $(-6, -10), (6, -1)$

9) $(5, 9), (-11, -3)$

10) $(3, -1), (1, -3)$

11) $(6, 0), (36, 72)$

12) $(8, 4), (3, -8)$

13) $(4, 2), (-5, -10)$

14) $(-8, 10), (4, 40)$

15) $(8, 4), (-10, -20)$

16) $(-16, -4), (32, 16)$

17) $(6, 10), (-10, -20)$

18) $(-5, 4), (7, 9)$

Find the midpoint of the line segment with the given endpoints.

1) $(-4, -4), (8, 4)$

2) $(20, 4), (-4, 4)$

3) $(6, -2), (2, 10)$

4) $(-6, -5), (2, 1)$

5) $(3, -2), (5, -2)$

6) $(-10, -4), (6, -2)$

7) $(4, 1), (-4, 9)$

8) $(-5, 6), (-5, 2)$

9) $(-8, 8), (4, -2)$

10) $(1, 7), (5, -1)$

11) $(-9, 5), (5, 3)$

12) $(7, 10), (-3, -6)$

13) $(-8, 14), (-8, 2)$

14) $(16, 7), (6, -3)$

15) $(5, 18), (-3, 12)$

16) $(-18, -1), (-10, 7)$

17) $(13, 9), (33, 27)$

18) $(-16, -22), (36, -2)$

Answers of Worksheets

Distributive and Simplifying Expressions

1) $4x - 5$
2) $5 + 7x$
3) $6x - 15$
4) $-3x^2 - 12x$
5) $5x^2 - 2x$
6) $2x + 15y$
7) $11x - 6y$
8) $6x - 1$
9) $-8x + 21$
10) $-2x - 5$

11) $2x - 2y$
12) $-12.5x^3$
13) $4x^2 - 4$
14) $14x^2 + 12$
15) $-4x + 2$
16) $-5x^2 - 4x$
17) $2x - 54$
18) $60x - 20$
19) $-6x - 18$
20) $8x$

21) $-4x^2 + 7x$
22) $5x + y + 5xy$
23) $-3x - 12$
24) $-3x + 4$
25) $3x + 4y - 5$
26) $-12x^2 - 2$
27) $8x + 8y$
28) $-6x - 14$

Factoring Expressions

1) $3(5x - 4)$
2) $3(x - 4)$
3) $3(x - 8)$
4) $6(b - 5)$
5) $4a(a - 6)$
6) $2y(x - 5)$
7) $5x(xy + 3)$
8) $a(a - 8 + 7b)$
9) $2a(a + b)$
10) $4(x + 5)$

11) $12x(2 - 3y)$
12) $2(4x - 3)$
13) $\frac{1}{4}(x - 3y)$
14) $7x(y - \frac{2}{3})$
15) $4(ab + 3c)$
16) $\frac{1}{5}(x - 4)$
17) $6x(2 - 3y)$
18) $x(x + 8)$
19) $4(x^2 - 3y)$

20) $x(4x^2 + 3y + x)$
21) $7(3x - 2)$
22) $20(b - 3c + d)$
23) $8a(3b - c)$
24) $(x - y)(a - 3)$
25) $(3x + 4)(a + 3)$
26) $x(x - 15)$
27) $7x^2(x - 2)$
28) $5x(x - 12y)$

Evaluate One Variable Expressions

1) 12
2) -15
3) -10
4) 4
5) 8
6) -4
7) 11
8) 6
9) -21
10) 6
11) 58
12) 26
13) 4
14) 5
15) 53
16) 16
17) 36
18) 6
19) 4
20) -33
21) 13
22) 4
23) -7
24) -14

| 25) 11 | 26) 27 | 27) 20 | 28) 18 |

Evaluate Two Variable Expressions

1) 33	9) 27	17) −7	24) 10
2) 10	10) 3	18) 34	25) −3
3) 26	11) 18	19) 8	26) 186
4) 7	12) 20	20) 70	27) −1
5) −20	13) 17	21) 25	28) $\frac{30}{29}$
6) 35	14) 78	22) 80	
7) 16	15) 12	23) 36	
8) 7	16) 90		

Graphing Lines Using Line Equation

1) $y = 2x - 5$

2) $y = -2x + 3$

3) $x - y = 0$

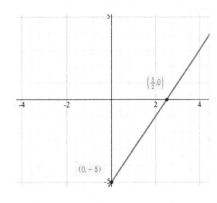

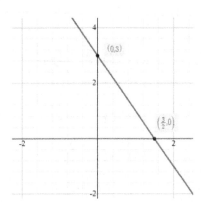

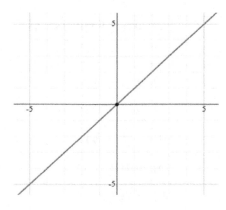

4) $x + y = 3$

5) $5x + 3y = -2$

6) $y - 3x + 2 = 0$

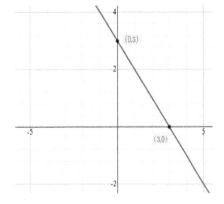

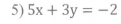
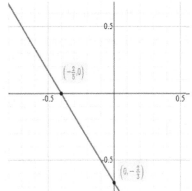

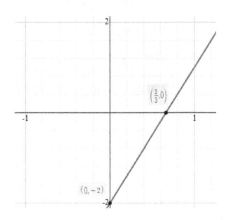

One Step Equations

1) $x = 112$
2) $x = -9$
3) $x = 9$
4) $x = -11$
5) $x = -14$
6) $x = -13$
7) x = 75
8) x = -45
9) x = 42
10) x = 1
11) x = 8
12) x = -13
13) x = -13
14) x = -2
15) x = 7
16) x = -10
17) $x = 4$
18) x = 11
19) x = 33
20) x = 5
21) x = 21
22) x = 32
23) $x = -3$
24) -8
25) -4
26) -72

Two Steps Equations

1) $x = 4$
2) $x = -6$
3) $x = 2$
4) $x = -5$
5) $x = -3$
6) $x = 2$
7) $x = -4$
8) $x = -1$
9) $x = 2$
10) $x = 5$
11) $x = -25$
12) $x = -7$
13) $x = 36$
14) $x = -1$
15) $x = -41$
16) $x = -1$
17) $x = -4$
18) $x = -30$
19) $x = 1$
20) $x = -20$
21) $x = -38$
22) $x = 3$
23) $x = 12$
24) $x = 36$

Multi Steps Equations

1) $x = 1$
2) $x = 2$
3) $x = 4$
4) $x = 5$
5) $x = -3$
6) $x = 13$
7) $x = 8$
8) $x = 2$
9) $x = -4$
10) $x = -24$
11) $x = -0.1$
12) $x = 7$
13) $x = 8$
14) $x = 3$
15) $x = -4$
16) $x = 5$
17) $x = 5$
18) $x = -3/5$
19) $x = -3$
20) $x = -3$
21) $x = -4$
22) $x = 2$
23) $x = 5$
24) $x = 3$

Graphing Linear Inequalities

1) $y > 2x - 3$ 2) $y < x + 3$ 3) $y \leq -3x - 8$

4) $3y \geq 6 + 3x$ 5) $-3y < x - 12$ 6) $2y \geq -8x + 4$

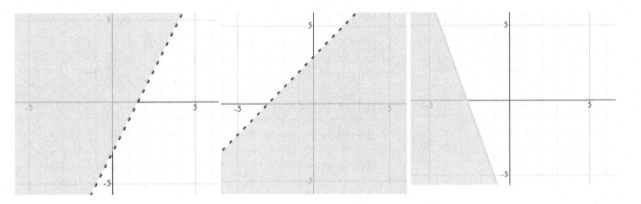
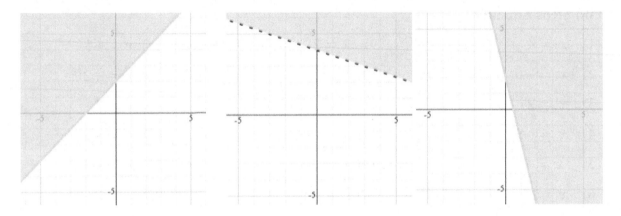

One Step Inequality

1) $x < 2$ 9) $x < 11$ 17) $x \leq -6$

2) $x \geq -21$ 10) $x > -36$ 18) $x \leq 9$

3) $x \leq 12$ 11) $x < 12$ 19) $x < 11$

4) $x < 7$ 12) $x \leq 32$ 20) $x \geq -18$

5) $x \geq -24$ 13) $x \geq 1$ 21) $x \leq 2$

6) $x \geq -4$ 14) $x \geq -39$ 22) $x \geq 2$

7) $x \leq \frac{16}{3}$ 15) $x \geq -1$ 23) $x < 3$

8) $x > -6$ 16) $x \geq -20$ 24) $x \leq 7$

Two Steps Inequality

1) $x \leq 5$ 3) $x \leq 0.5$ 5) $x \geq 4$

2) $x \leq 4$ 4) $x \geq 4$ 6) $x \leq 7$

7) $x \leq 2$

8) $x \leq 2$

9) $x > 11$

10) $x \leq 16$

11) $x \geq 11$

12) $x < 6$

13) $x \leq 5$

14) $x < 13$

15) $x \geq 11$

16) $x < 5$

17) $x < 4$

18) $x < 8$

19) $x > 7$

20) $x < -27$

Multi Steps Inequality

1) $x > 3$

2) $x \geq 5$

3) $x > 5$

4) $x > \frac{4}{3}$

5) $x < -12$

6) $x < -1.5$

7) $x < -4$

8) $x < -17$

9) $x < 3$

10) $x \geq 1.6$

11) $x \geq -13$

12) $x > 8.5$

13) $x < -3$

14) $x \geq 4$

15) $x < -3$

16) $x \leq -2$

17) $x < \frac{1}{5}$

18) $x > 1$

19) $x > -56$

20) $x \geq 4$

Systems of Equations

1) $x = 1, y = 2$

2) $x = 3, y = 2$

3) $x = -\frac{33}{2}$

4) $x = -\frac{1}{5}, y = \frac{26}{5}$

5) $x = -4, y = -3$

6) $x = 1, y = -2$

7) $x = 1, y = 2$

8) $x = \frac{9}{5}, y = \frac{17}{5}$

9) $x = -110, y = -85$

10) $x = -\frac{15}{4}, y = \frac{29}{4}$

11) $x = \frac{5}{3}, y = \frac{2}{3}$

12) $x = \frac{29}{4}, y = -\frac{15}{4}$

13) $x = \frac{33}{19}, y = -\frac{15}{19}$

14) $x = -2, y = -4$

15) $x = -\frac{2}{7}, y = \frac{27}{14}$

16) $x = \frac{91}{17}, y = -\frac{132}{17}$

Systems of Equations Word Problems

1) 7

2) $16

3) 8

4) 1

5) $21

6) 24

7) 43

8) 18, 7

9) 11 meters

10) 18

Finding Distance of Two Points

1) 10

2) 20

3) 60

4) 26

5) 15

6) 5

7) 17

8) 15

9) 20

10) $2\sqrt{2}$

11) 78

12) 13

13) 15   15) 30   17) 34

14) $6\sqrt{29}$ 16) 52   18) 13

Finding Midpoint

1) $(2, 0)$ 7) $(0, 5)$ 13) $(-8, 8)$

2) $(8, 4)$ 8) $(-5, 4)$ 14) $(11, 2)$

3) $(4, 2)$ 9) $(-2, 3)$ 15) $(1, 15)$

4) $(-2, -4)$ 10) $(3, 3)$ 16) $(-14, 3)$

5) $(4, -2)$ 11) $(-2, 4)$ 17) $(23, 18)$

6) $(-2, -3)$ 12) $(2, 2)$ 18) $(10, -12)$

Chapter 9 :

Linear Functions

Relation and Functions

Determine whether each relation is a function. Then state the domain and range of each relation.

1)

Function:

....

Domain:

....

Range:

....

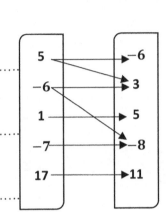

2)

Function:

....

Domain:

....

Range:

....

x	y
7	6
3	4
−8	−9
8	−9
−11	2

3)

Function:

....

Domain:

....

Range:

....

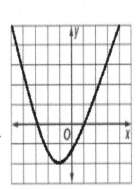

4) $\{(2,-2),(7,-6),(9,9),(8,1),(7,4)\}$

Function:

....

Domain:

....

Range:

....

5)

Function:

....

Domain:

....

Range:

....

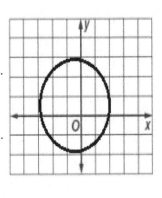

6)

Function:

....

Domain:

....

Range:

....

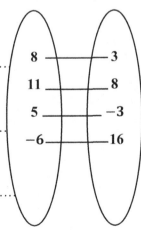

Slope form

Write the slope-intercept form of the equation of each line.

1) $6x + 7y = 14$

2) $8x + 24y = 6$

3) $14x + 2y = -18$

4) $-7x + 11y = 5$

5) $5x - 4y = 7$

6) $-21x + 3y = 6$

7) $2x + y = 0$

8) $5x - 7y = -9$

9) $-13.5x + 27y = 54$

10) $-3x + \frac{2}{3}y = 18$

11) $10x + y = -120$

12) $6x = -72y - 54$

13) $4.5x = 9y + 9$

14) $10x = -\frac{10}{8}y + 50$

Slope and Y-Intercept

Find the slope and y-intercept of each equation.

1) $y = \frac{1}{8}x + 5$

2) $y = 10x + 7$

3) $x - 6y = 18$

4) $y = 10x + 22$

5) $y = 9$

6) $y = -2x + 3$

7) $x = -15$

8) $y = 7x$

9) $y - 5 = 8(x + 1)$

10) $x = -\frac{11}{8}y - \frac{1}{6}$

Slope and One Point

Find a Point-Slope equation for a line containing the given point and having the given slope.

1) $m = -2, (1, -1)$

2) $m = 3, (1, 2)$

3) $m = -2, (-1, -5)$

4) $m = 2, (6, 4)$

5) $m = 5, (2, 4)$

6) $m = \frac{3}{2}, (4, 5)$

7) $m = 0, (-4, -5)$

8) $m = 2, (1, -3)$

9) $m = 1, (0, 3)$

10) $m = \frac{3}{4}, (-2, -5)$

11) $m = -3, (1, -1)$

12) $m = -2, (2, -1)$

13) $m = 5, (1, 0)$

14) $m =$ undefined, $(8, -8)$

15) $m = -\frac{1}{8}, (8, 4)$

16) $m = \frac{1}{4}, (3, 2)$

17) $m = -8, (2, 4)$

18) $m = 6, (-2, -4)$

19) $m = \frac{1}{3}, (3, 1)$

20) $m = \frac{-4}{9}, (0, -3)$

21) $m = \frac{1}{4}, (4, 4)$

22) $m = -5, (0, -1)$

23) $m = 0, (0.9, -3)$

24) $m = -\frac{2}{5}, (5, -2)$

25) $m = 0, (-2, 17)$

26) $m =$ Undefined, $(-9, -3)$

Slope of Two Points

Write the slope-intercept form of the equation of the line through the given points.

1) $(2, 0), (-2, 10)$

2) $(-2, 6), (10, 12)$

3) $(-10, 2), (-2, 10)$

4) $(2, -3), (-9, 8)$

5) $(5, 0), (3, 1)$

6) $(9, -1), (-1, 9)$

7) $(-5, 3), (-6, 1)$

8) $(-7, -2), (1, 0)$

9) $(-5, -5), (3, 3)$

10) $(-1, 9), (-1, -5)$

11) $(-2, 7), (1, 7)$

12) $(1, -5), (4, -4)$

13) $(6, -9), (-3, 0)$

14) $(1, -4), (7, 4)$

15) $(-9, 5), (-3, -1)$

16) $(9, 5), (5, 1)$

17) $(10, -7), (2, -6)$

18) $(-5, -9), (-7, 2)$

19) $(7, 4), (3, 1)$

20) $(-1, -1), (9, 2)$

21) $(-8, 8), (8, 2)$

22) $(9, 2), (5, 11)$

23) $(16, 4), (18, 6)$

24) $(-4, -10), (-10, -16)$

Equation of Parallel and Perpendicular lines

Write the slope-intercept form of the equation of the line described.

1) Through: $(-5, 4)$, parallel to $y = 4x + 10$

2) Through: $(-2, 1)$, parallel to $y = -7x$

3) Through: $(-10, -2)$, perpendecular to $y = \frac{1}{2}x + 8$

4) Through: $(6, -2)$, parallel to $y = -5x + 13$

5) Through: $(-7, 4)$, parallel to $y = \frac{3}{7}x - 6$

6) Through: $(2, 0)$, perpendecular to $y = -\frac{1}{5}x + 8$

7) Through: $(4, -7)$, perpendecular to $y = -6x - 10$

8) Through: $(-5, 1)$, perpendecular to $y = -\frac{1}{8}x + 3$

9) Through: $(-1, -2)$, parallel to $2y + 4x = 9$

10) Through: $(1, 10)$, parallel to $y = \frac{1}{10}x - 5$

11) Through: $(5, -5)$, parallel to $y = 9$

12) Through: $(7, 2)$, perpendecular to $y = \frac{5}{2}x + 3$

13) Through: $(0, -4)$, perpendecular to $3y - x = 11$

14) Through: $(3, 5)$, parallel to $3y + x = 5\frac{3}{4}$

15) Through: $(1, 1)$, perpendecular to $y = 5x + 12$

16) Through: $(-2, -4)$, parallel to $6y - x = 11$

17) Through: $(-1, -1)$, perpendecular to $y = x + \frac{1}{2}$

18) Through: $(-6, 0)$, perpendecular to $5y - 2x - 9 = 0$

Answers of Worksheets

Relation and Functions

1) No, $D_f = \{5, -6, 1, -7, 17\}$, $R_f = \{-6, 3, 5, -8, 11\}$

2) Yes, $D_f = \{7, 3, -8, 8, -11\}$, $R_f = \{6, 4, -9, 2\}$

3) Yes, $D_f = (-\infty, \infty)$, $R_f = \{-2, \infty)$

4) No, $D_f = \{2, 7, 9, 8, 7\}$, $R_f = \{-2, -6, 9, 1, 4\}$

5) No, $D_f = [-3, 2]$, $R_f = [-2, 3]$

6) Yes, $D_f = \{8, 11, 5, -6\}$, $R_f = \{2, 8, -3, 16\}$

Slope form

1) $y = -\frac{6}{7}x + 2$

2) $y = -\frac{1}{3}x + \frac{1}{4}$

3) $y = -7x - 9$

4) $y = \frac{7}{11}x + \frac{5}{11}$

5) $y = \frac{5}{4}x - \frac{7}{4}$

6) $y = 7x + 2$

7) $y = -2x$

8) $y = \frac{5}{7}x + \frac{9}{7}$

9) $y = 0.5x + 2$

10) $y = 4.5x + 27$

11) $y = -10x - 120$

12) $y = -\frac{1}{12}x - \frac{3}{4}$

13) $y = 0.5x - 1$

14) $y = -8x + 40$

Slope and Y-Intercept

1) $m = \frac{1}{8}, b = 5$

2) $m = 10, b = 7$

3) $m = \frac{1}{6}, b = -3$

4) $m = 10, b = 22$

5) $m = 0, b = 9$

6) $m = -2, b = 3$

7) $m = undefind$, $b: no\ intercept$

8) $m = 7, b = 0$

9) $m = 8, b = 13$

10) $m = -\frac{8}{11}, b = -\frac{1}{6}$

Slope and One Point

1) $y = -2x + 1$

2) $y = 3x - 1$

3) $y = -2x - 7$

4) $y = 2x - 8$

5) $y = 5x - 6$

6) $y = \frac{3}{2}x - 1$

7) $y = -5$

8) $y = 2x - 5$

9) $y = x + 3$

10) $y = \frac{3}{4}x - \frac{7}{2}$

11) $y = -3x + 2$

12) $y = -2x + 3$

13) $y = 5x$

14) $x = 8$

15) $y = -\frac{1}{8}x + 5$

16) $y = \frac{1}{4}x + \frac{5}{4}$

17) $y = -8x + 20$

18) $y = 6x + 8$

19) $y = \frac{1}{3}x$

20) $y = -\frac{4}{9}x - 3$

21) $y = \frac{1}{4}x + 3$

22) $y = -5x - 1$

23) $y = -3$

24) $y = -\frac{2}{5}x$

25) $y = 17$

26) $x = -9$

Slope of Two Points

1) $y = -\frac{5}{2}x + 5$

2) $y = \frac{1}{2}x + 7$

3) $y = x + 12$

4) $y = -x - 1$

5) $y = -\frac{1}{2}x + \frac{5}{2}$

6) $y = -x + 8$

7) $y = 2x + 13$

8) $y = \frac{1}{4}x - \frac{1}{4}$

9) $y = x$

10) $x = -1$

11) $y = 7$

12) $y = \frac{1}{3}x - 5\frac{1}{3}$

13) $y = -x - 3$

14) $y = \frac{4}{3}x - 5\frac{1}{3}$

15) $y = -x - 4$

16) $y = x - 4$

17) $y = -\frac{1}{8}x - 5\frac{3}{4}$

18) $y = -5\frac{1}{2}x - 36\frac{1}{2}$

19) $y = \frac{3}{4}x - 1\frac{1}{4}$

20) $y = \frac{3}{10}x - \frac{7}{10}$

21) $y = -\frac{3}{8}x + 5$

22) $y = -\frac{9}{4}x + 22\frac{1}{4}$

23) $y = x - 12$

24) $y = x - 6$

Equation of Parallel and Perpendicular lines

1) $y = 4x + 24$

2) $y = -7x - 13$

3) $y = -2x - 22$

4) $y = -5x + 28$

5) $y = \frac{3}{7}x + 7$

6) $y = 5x - 10$

7) $y = \frac{1}{6}x - 7\frac{2}{3}$

8) $y = 8x + 41$

9) $y = -2x - 4$

10) $y = \frac{1}{10}x + 9\frac{9}{10}$

11) $y = -5$

12) $y = -\frac{2}{5}x + 4\frac{4}{5}$

13) $y = -3x - 4$

14) $y = -\frac{1}{3}x + 6$

15) $y = -\frac{1}{5}x + 1\frac{1}{5}$

16) $y = \frac{1}{6}x - 3\frac{2}{3}$

17) $y = -x - 2$

18) $y = -\frac{5}{2}x - 15$

Chapter 10 : Statistics and probability

Mean, Median, Mode, and Range of the Given Data

Find the mean, median, mode(s), and range of the following data.

1) 10, 2, 38, 23, 47, 23, 21

Mean: __, Median: __, Mode: __, Range: __

2) 12, 26, 26, 38, 30, 20

Mean: __, Median: __, Mode: __, Range: __

3) 41, 24, 49, 11, 45, 27, 35, 19, 24

Mean: __, Median: __, Mode: __, Range: __

4) 25, 11, 1, 15, 25, 18

Mean: __, Median: __, Mode: __, Range: __

5) 24, 14, 14, 17, 23, 15, 14, 29, 29, 8

Mean: __, Median: __, Mode: __, Range: __

6) 7, 14, 19, 11, 8, 19, 8, 15

Mean: __, Median: __, Mode: __, Range: __

7) 29, 28, 66, 76, 14, 44, 18, 44, 22, 44

Mean: __, Median: __, Mode: __, Range: __

8) 35, 35, 57, 78, 59

Mean: __, Median: __, Mode: __, Range: __

9) 16, 16, 29, 46, 54

Mean: __, Median: __, Mode: __, Range: __

10) 13, 9, 3, 3, 5, 6, 7

Mean: __, Median: __, Mode: __, Range: __

11) 4, 12, 4, 6, 1, 8

Mean: __, Median: __, Mode: __, Range: __

12) 8, 9, 15, 15, 17, 17, 17

Mean: __, Median: __, Mode: __, Range: __

13) 7, 7, 1, 16, 1, 7, 19

Mean: __, Median: __, Mode: __, Range: __

14) 13, 17, 10, 12, 12, 18, 15, 19

Mean: __, Median: __, Mode: __, Range: __

15) 29, 14, 30, 19, 29

Mean: __, Median: __, Mode: __, Range: __

16) 6, 6, 16, 18, 15, 22, 37

Mean: __, Median: __, Mode: __, Range: __

17) 25, 11, 14, 25, 18, 13, 7, 5

Mean: __, Median: __, Mode: __, Range: __

18) 55, 34, 34, 48, 85, 7

Mean: __, Median: __, Mode: __, Range: __

19) 54, 28, 28, 65, 5, 8

Mean: __, Median: __, Mode: __, Range: __

20) 77, 94, 25, 24, 11, 77, 19

Mean: __, Median: __, Mode: __, Range: __

Box and Whisker Plot

1) Draw a box and whisker plot for the data set:

   16, 11, 14, 12, 14, 12, 16, 16, 20

2) The box-and-whisker plot below represents the math test scores of 20 students.

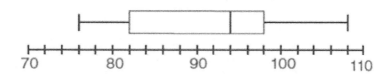

   A. What percentage of the test scores are less than 82?

   B. Which interval contains exactly 50% of the grades?

   C. What is the range of the data?

   D. What do the scores 76, 94, and 108 represent?

   E. What is the value of the lower and the upper quartile?

   F. What is the median score?

Histogram

Create a histogram for the set of data.

Math Test Score out of 100 points.

58	74	63	80	83	65	70	86	67	54
81	73	82	75	71	56	87	66	74	72
84	55	76	73	67	85	69	68	52	87

Frequency Table	
Interval	**Number of Values**

Dot plots

The ages of students in a Math class are given below.

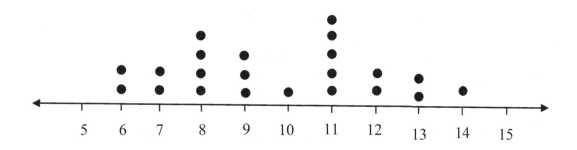

1) What is the total number of students in math class?

2) How many students are at least 12 years old?

3) Which age(s) has the most students?

4) Which age(s) has the fewest student?

5) Determine the median of the data.

6) Determine the range of the data.

7) Determine the mode of the data.

Scatter Plots

A person charges an hourly rate for his services based on the number of hours a job takes.

Hours	Rate
1	$24.5
2	$22
3	$21
4	$19.50

Hours	Rate
5	$19
6	$17.50
7	$17
8	$16.5

1) Draw a scatter plot for this data.

2) Does the data have positive or negative correlation?

3) Sketch the line that best fits the data.

4) Find the slope of the line.

5) Write the equation of the line using slope-intercept form.

6) Using your prediction equation: If a job takes 10 hours, what would be the hourly rate?

Pie Graph

60 people were survey on their favorite ice cream. The pie graph is made according to their responses. Answer following questions based on the Pie graph.

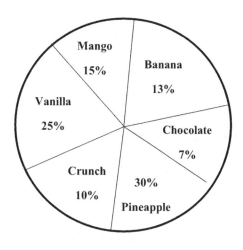

1) How many people like to eat Mango ice cream? _____

2) Approximately, which two ice creams did about half the people like the

   best? _____

3) How many people said either mango or crunch ice cream was their favorite?

   _____

4) How many people would like to have crunch ice cream? _____

5) Which ice cream is the favorite choice of 15 people? _____

Probability

1) A jar contains 16 caramels, 5 mints and 19 dark chocolates. What is the probability of selecting a mint?

2) If you were to roll the dice one time what is the probability it will NOT land on a 4?

3) A die has sides are numbered 1 to 6. If the cube is thrown once, what is the probability of rolling a 5?

4) The sides of number cube have the numbers 4, 6, 8, 4, 6, and 8. If the cube is thrown once, what is the probability of rolling a 6?

5) Your friend asks you to think of a number from ten to twenty. What is the probability that his number will be 15?

6) A person has 8 coins in their pocket. 2 dime, 3 pennies, 2 quarter, and a nickel. If a person randomly picks one coin out of their pocket. What would the probability be that they get a penny?

7) What is the probability of drawing an odd numbered card from a standard deck of shuffled cards (Ace is one)?

8) 32 students apply to go on a school trip. Three students are selected at random. what is the probability of selecting 4 students?

Answers of Worksheets

Mean, Median, Mode, and Range of the Given Data

1) mean: 23.43, median: 23, mode: 23, range: 45.

2) mean: 25.33, median: 26, mode: 26, range: 26.

3) mean: 30.56, median: 27, mode: 24, range: 38.

4) mean: 15.83, median: 16.5, mode: 25, range: 24.

5) mean: 18.7, median: 16, mode: 14, range: 21.

6) mean: 12.63, median: 12.5, mode: 19, 8, range: 12.

7) mean: 38.5, median: 36.5, mode: 44, range: 62.

8) mean: 52.8, median: 57, mode: 35, range: 43.

9) mean: 32.2, median: 29, mode: 16, range: 38.

10) mean: 6.57, median: 6, mode: 3, range: 10.

11) mean: 5.83, median: 5, mode: 4, range: 11.

12) mean: 14, median: 15, mode: 17, range: 9.

13) mean: 8.29, median: 7, mode: 7, range: 18.

14) mean: 14.5, median: 14, mode: 12, range: 9.

15) mean: 24.2, median: 29, mode: 29, range: 16.

16) mean: 17.14, median: 16, mode: 6, range: 31.

17) mean: 14.75, median: 13.5, mode: 25, range: 20.

18) mean: 43.83, median: 41, mode: 34, range: 78.

19) mean: 31.33, median: 28, mode: 28, range: 60.

20) mean: 46.71, median: 25, mode: 77, range: 83.

Box and Whisker Plot

1)

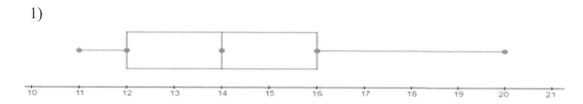

2)

A. 25%

B. 94

C. 32

D. Minimum, Median, and Maximum.

E. Lower (Q_1) is 82 and upper (Q_3)is 98.   F. 94

Histogram

Frequency Table	
Interval	**Number of Values**
52-57	4
58-63	2
64-69	6
70-75	8
76-81	3
82-87	7

Histogram

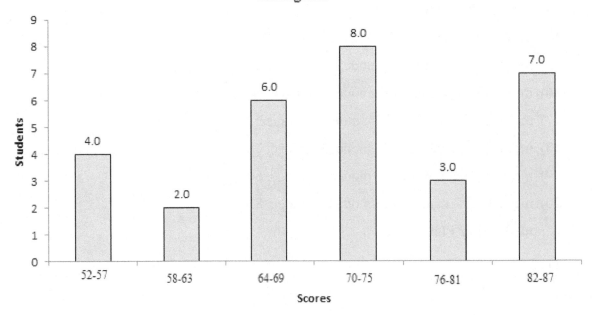

Dot plots

1) 22   5) 2

2) 5   6) 4

3) 11   7) 2

4) 10 and 14

Scatter Plots

1)

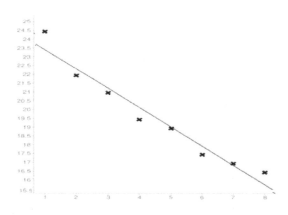

2) Negative correlation

3) ----

4) Slope(m)= -1

5) $y = -x + 24.5$

6) 14.5

Pie Graph

1) 9

2) Vanilla and pineapple

3) 15

4) 6

5) Vanilla

Probability

1) $\frac{1}{8}$

2) $\frac{5}{6}$

3) $\frac{1}{6}$

4) $\frac{1}{3}$

5) $\frac{1}{10}$

6) $\frac{3}{8}$

7) $\frac{5}{13}$

8) $\frac{1}{8}$

Chapter 11 :

PSSA Mathematics

Test Review

Grade 7 PSSA Mathematics Formula Sheet

Formulas that you may need to work questions on this test are found below. You may refer to this page at any time during the mathematics test.

Triangle

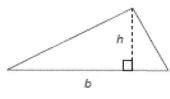

$$A = \frac{1}{2}bh$$

Rectangle

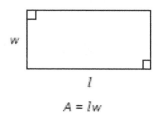

$$A = lw$$

Square

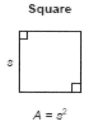

$$A = s^2$$

Parallelogram

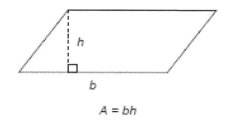

$$A = bh$$

Trapezoid

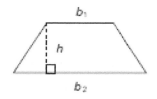

$$A = \frac{1}{2}h(b_1 + b_2)$$

Rectangular Prism

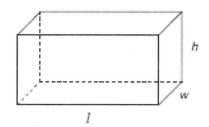

$$V = lwh \qquad SA = 2lw + 2lh + 2wh$$

Cube

$$V = s \cdot s \cdot s \qquad SA = 6s^2$$

Triangular Prism

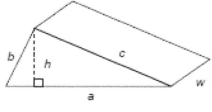

$$SA = ah + aw + bw + cw$$

The Pennsylvania System

of School Assessment

PSSA Practice Test 1

Mathematics

GRADE 7

❖ **20 Questions**

❖ **Calculators are permitted for this practice test.**

Pennsylvania Department of Education Bureau of Curriculum, Assessment, and Instruction— *Month Year*

1) Peter paid for 5 sandwiches.

- Each sandwich cost 9.85.

- He paid for 4 bags of fries that each cost $2.13.

Which equation can be used to determine the total amount, y, Peter paid?

A. $y = 5(9.85) + 4(2.13)x$

B. $y = (9.85 + 2.13)x$

C. $y = 5(9.85) + 4(2.13)$

D. $y = 9.85x + 4(2.13)$

2) The circumference of a circle is $14\,\pi$ centimeters. What is the area of the circle in terms of π?

A. $14\,\pi$

B. $49\,\pi$

C. $149\,\pi$

D. $98\,\pi$

3) If 15% of x is 60, what is 35% of x?

A. 140

B. 14.40

C. 40.14

D. 4.44

4) If all variables are positive, find the square root of $\frac{49x^9y^3}{36xy}$?

A. $\frac{3}{7}x^8y$

B. $\frac{6y}{7x^4}$

C. $\frac{7}{6}x^4y$

D. $7x^8y^2$

5) What is the volume of rectangular prism when the two triangular prisms below are stuck together?

A. $900\ in^3$

B. $450\ in^3$

C. $1,350\ in^3$

D. $36\ in^3$

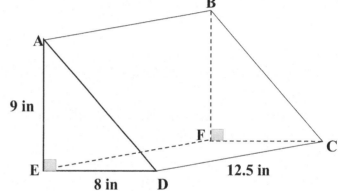

6) A school has 384 students and 21 chemistry teachers and 16 physics teachers. What is the ratio between the number of physics teachers and the number of students at the school?

A. $\frac{1}{24}$

B. $\frac{5}{12}$

C. $\frac{1}{16}$

D. $\frac{16}{384}$

7) Which number line Shows the solution to the inequality $-4x + 7 < -5$?

A.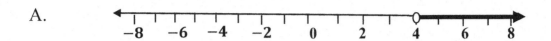

B.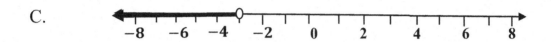

C.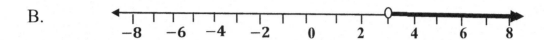

D.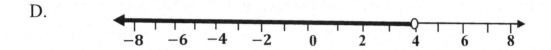

8) The medals won by United States, Australia and Spain during a basketball competition are shown in the table below:

United States	Australia	Spain
18	16	26

Out of the medals won by these three countries, what percentage of medals did the United States win?

A. 9%

B. 30%

C. 70%

D. 39%

9) Arsan has $11 to spend on school supplies. The following table shows the price of each item in the school store. No sale tax is charged on these items.

Which the combination of items can Arsan buy with his $11?

A. 3 Notebooks and 5 Pens

B. 3 Folders and 6 Erasers

C. 5 Notebooks and 4 Folders

D. 7 Erasers and 3 Pens

Item	Price
Notebook	$3.20
Pen	$0.90
Eraser	$ 0.85
Folder	$2.24

10) Which number represents the probability of an event that is very likely to occur?

A. 0.18

B. 1.1

C. 0.82

D. 0.02

11) The ratio of boys to girls in Maria Club is the same as the ratio of boys to girls in Hudson Club. There are 16 boys and 64 girls in Maria Club. There are 22 boys in Hudson Club. How many girls are in Hudson Club?

A. 24

B. 48

C. 80

D. 88

12) A girl in State A spent $45 before a 3.4% sales tax and a girl in State B spent $28 before an 4.5% sales tax. How much more money did the girl from State A spend than the girl from State B after sales tax was applied? Round to the nearest hundredth.

A. 17.27

B. 17.17

C. 19.29

D. 29.19

13) On average, Simone drinks $\frac{3}{8}$ of a 16-ounce glass of coffee in $\frac{1}{5}$ hour. How much coffee does she drink in an hour?

A. 45 ounces

B. 15 ounces

C. 30 ounces

D. 40 ounces

14) Line P, R, and S intersect each other, as shown in below diagram. Based on the angle measures, what is the value of θ?

A. 152°

B. 132°

C. 118°

D. 110°

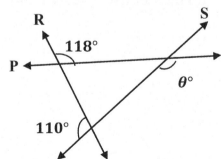

15) The temperature is shown in the table below, on each of day in the week for a city in February. What is the mean temperature, in the city for that week?

A. −13

B. −13.2

C. −3

D. −3.4

Day	Temperature (°F)
Monday	−25
Tuesday	−28
Wednesday	−21
Thursday	2
Friday	19
Saturday	−6
Sunday	38

16) James has his own lawn mowing service. The maximum James charges to mow a lawn is $25. Which inequality represents the amount James could charge, P, to mow a lawn?

A. P ≤ 25

B. P = 25

C. P ≥ 25

D. P < 25

17) Which expression is represented by the model below?

A. (−7).(−4)

B. (−7).4

C. 4.(−7)

D. 4.7

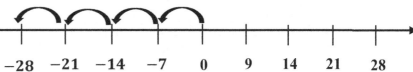

−28   −21   −14   −7   0   9   14   21   28

18) Which graph best represents the distance a car travels when going 30 miles per hour?

A.

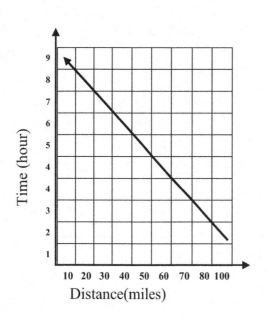

B.

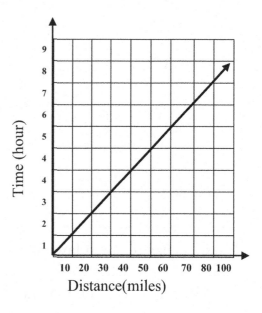

C.

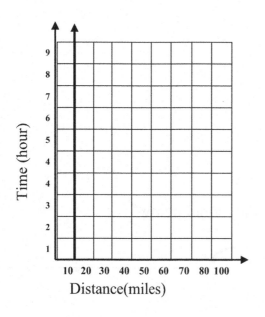

D.

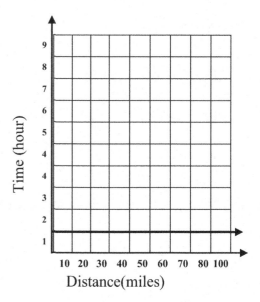

19) The dot plots show how many minutes per day do 7th grade study math after

school at two different schools on one day.

- Number of minuets study in school 1

- Number of minuets study in school 2

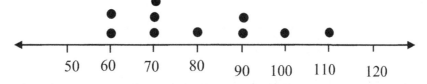

Which statement is supported by the information in the dot plots?

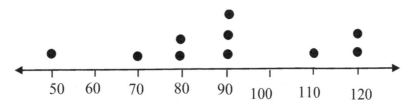

A. The mode of the data for School 2 is greater than the mode of the data for

School 1.

B. The mean of the data for School 1 is greater than the mean of the data for

School 2.

C. The median of the data for School 2 is smaller than the median of the data for

School 1.

D. The median and mean of the data for two schools are equal.

20) The table below shows the distance, y, a lion can travel in mile in x hour.

Time (x, hour)	Distance (y, mile)
4	128
8	256
12	384
16	512
20	640

Based on the information in the table, which equation can be used to model the

relationship between x and y?

A. $y = x + 4$

B. $y = 4x$

C. $y = x + 1280$

D. $y = 32x$

The Pennsylvania System

of School Assessment

PSSA Practice Test 2

Mathematics

GRADE 7

❖ **20 Questions**

❖ **Calculators are permitted for this practice test.**

Pennsylvania Department of Education Bureau of Curriculum, Assessment, and Instruction— *Month Year*

1) What is the decimal equivalent of the fraction $\frac{34}{11}$?

    A. $\overline{3.09}$

    B. $3.0\overline{90}$

    C. $3.\overline{09}$

    D. 3.09

2) Thomas is shareholder of a company. The price of stock is $75.85 on the morning of day 1. Thomas records the change in the price of the stock in the chart below at the end of each day, but some information is missing.

Day	Change in Price ($)
1	+ 0.52
2	+0.66
3	
4	−0.7
5	

The change in the price for day 3 is $\frac{2}{5}$ of the change in the price for day 4. At the end of day 5, the price of Thomas's stock is $76.92. What is the change, in dollars, in the price of the stock for day 5?

    A. −0.68

    B. 0.87

    C. 0.82

    D. 1.06

3) Kevin adds $\frac{5}{8}$ cups of sugar into a mixture every $\frac{1}{2}$ hour. What is the rate, in cups per minute, at which Kevin adds sugar to the mixture?

A. $\frac{1}{48}$

B. $2\frac{1}{3}$

C. $\frac{1}{18}$

D. $\frac{1}{24}$

4) Multiply: $2\frac{4}{15} \times \frac{-7}{15}$

A. $1\frac{13}{15}$

B. $-1\frac{2}{15}$

C. $-1\frac{13}{225}$

D. $-3\frac{13}{15}$

5) Brendan charges $51 per hour plus $68 to enter data. He accepted a project for no more than $730. Which inequality can be used to determine all the possible numbers of hours (x) it took the man to enter the data?

A. $51x + 68 \le 730$

B. $51x + 68 > 730$

C. $68x + 51 < 730$

D. $68x + 51 \ge 730$

6) Use the coordinate grid below to answer the question. What is the circumference of the circle? ($\pi = 3.14$)

A. 100.52

B. 50.24

C. 13.25

D. 25.13

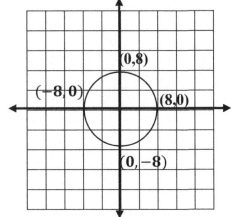

7) The temperature is 8° F. As a cold front move in, the temperature drops 3° F per half hour. What is the temperature at the end of 3 hours?

A. 2°F

B. 26°F

C. −1°F

D. −10°F

8) A printer originally cost h dollars, including tax. Eddy purchased the printer when it was on sale for 43% off its original cost. Which of the following expressions represents the final cost, in dollars, of the printer Eddy purchased?

A. $h + 0.57$

B. $h - 0.43$

C. $0.57h$

D. $0.43h$

9) Triangle PRS is shown on the grid below:

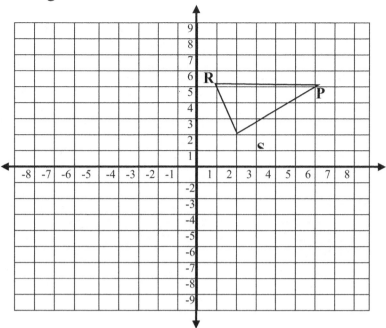

If triangle PRS is reflected across the y-axis to form triangle P′R′S′, which ordered pair represents the coordinates of P′?

    A. $(-1, -5)$

    B. $(1, -5)$

    C. $(-1, 5)$

    D. $(5, -1)$

10) What is the solution set for the inequality $-5x + 12 > -8$?

    A. $x > 4$

    B. $x < 4$

    C. $x > -4$

    D. $x < -4$

11) The store manager spent $11,840 to buy a new freezer and 25 tables. The total purchase is represented by this equation, where v stands for the value of each table purchased: $25v + 2,340 = 11,840$

What was the cost of each table that the manager purchased?

A. $308

B. $320

C. $580

D. $380

12) In a city, at 1:30 A.M., the temperature was $-7°F$. At 1:30 P.M., the temperature was $16°F$. Which expression represents the increase in temperature?

A. $-7 + 16$

B. $|-7 - 16|$

C. $|-7| - 16$

D. $-7 - |16|$

13) Angles α and β are complementary angles. Angles α and γ are supplementary angles. The degree measure of angle β is 125°. What is the measure of angle γ?

A. 35°

B. 52°

C. 42°

D. 48°

14) The bar graph shows a company's income and expenses over the last 5 years.

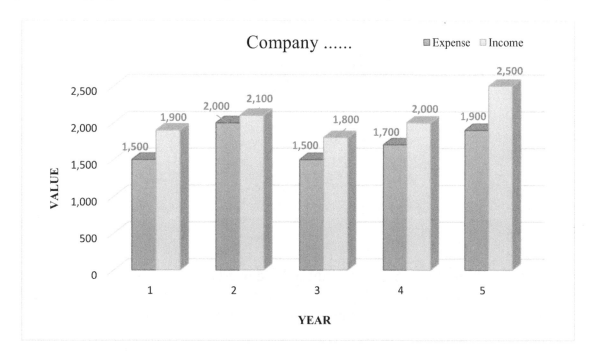

Which statement is supported by the information in the graph?

A. Expenses have increased $500 each year over the last 5 years.

B. The income in Year 5 was 20% more than the income in Year 1.

C. The combined income in Years 3, 4, and 5 was equal to the combined expenses in Years 2, 3, and 4.

D. Expenses in the year 3 was more than half of the income in the year 4.

15) Which expression is equivalent to the $(4n - 7) - \frac{1}{2}(11 - 8n) + \frac{7}{2}$?

A. -9

B. $-8n - 9$

C. $8n - 9$

D. $8n + 9$

16) Patricia bought a bottle of 16-ounce balsamic vinegar for $12.14 She used 45% of the balsamic vinegar in two weeks. Which of the following is closest to the cost of the balsamic she used?

A. $5.56

B. $4.65

C. $5.46

D. $4.56

17) A scale drawing of triangle DEF that will be used on a wall is shown below. What is the perimeter, in meter, of the actual triangle used on the wall?

Scale: 1 cm: $5\frac{1}{5}$ m

A. $54\frac{3}{5}$

B. $57\frac{3}{5}$

C. $57\frac{1}{5}$

D. $54\frac{1}{5}$

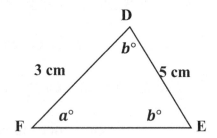

18) The ratio of boys to girls in Geometry class is 3 to 4. There are 24 girls in the class. What is the total number of students in Geometry class?

A. 18

B. 42

C. 48

D. 49

19) In a party people drink 88.71 liters of juices. There are approximately 29.57 milliliters in 1 fluid ounce. Which measurement is closest to the number of fluid ounces in 88.71 liters?

A. 0.003 fl oz

B. 1,600 fl oz

C. 3,600 fl oz

D. 3,000 fl oz

20) The angle measures of a triangle GBD are shown in the diagram. What is the value of ∠B?

A. 20°

B. 104°

C. 134°

D. 44°

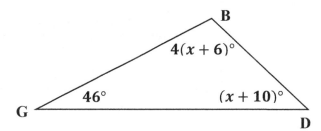

Chapter 12 : Answers and Explanations

Answer Key

Now, it's time to review your results to see where you went wrong and what areas you need to improve!

PSSA Math Practice Tests

Practice Test 1

1	C	11	D
2	B	12	A
3	A	13	C
4	C	14	B
5	A	15	C
6	A	16	A
7	B	17	C
8	B	18	B
9	D	19	A
10	C	20	D

Practice Test 2

1	C	11	D
2	B	12	B
3	A	13	A
4	C	14	D
5	A	15	C
6	B	16	C
7	D	17	C
8	C	18	B
9	C	19	D
10	B	20	B

PSSA Practice Test 1

Answers and Explanations

1) Answer: C

Let y be the total amount paid.

We have been given that Peter bought 5 sandwiches that each cost the $9.85. So, the cost of 5 sandwiches would be 5 (9.85). He paid for 4 bags of fries that each cost the $2.13. So, the cost of 4 bags would be 4(2.13)

Then, the total cost of sandwiches and fries would be $y = 5(9.85) + 4(2.13)$

2) Answer: B

Use the formula of circumference of circles.

Circumference = πd = 2π (r) = 14 π ⇒r=7

Radius of the circle is 7. Now, use the areas formula:

Area = πr^2 ⇒ Area = $\pi(7)^2$ ⇒ Area = 49 π

3) Answer: A

$0.15 \times x = 60 \rightarrow x = \frac{60}{0.15} = \frac{6,000}{15} = 400$

$35\% of\ 400 = 0.35 \times 400 = 140$

4) Answer: C

$\sqrt{\frac{49x^9y^3}{36xy}} = \sqrt{\frac{49}{36} \times \frac{x^9y^3}{xy}} = \sqrt{\frac{49}{36}x^8y^2} = \frac{7}{6}x^4y$

5) Answer: A

The volume of a triangular prism is the base times the height. $V = Bh$

Area of the base = $\frac{1}{2}b.h \rightarrow B = \frac{1}{2} \times 8 \times 9 = 36$

$V = B.h = 36 \times 12.5 = 450$ and we need two triangular prisms, then,

$2 \times 450 = 900$

6) Answer: A

16 physics teachers to 384 students are 16:384, 1:24

7) Answer: B

$-4x + 7 < -5$, add -7 to both sides $-4x < -12$ divide each term by -4

If an inequality is multiplied or divided by a negative number, you must change the direction of the inequality, then $x > 3$

8) Answer: B

Use percent formula: part $= \dfrac{\text{percent}}{100} \times$ whole

Whole $= 18 + 16 + 26 = 60$

$18 = \dfrac{\text{percent}}{100} \times 60 \Rightarrow 18 = \dfrac{\text{percent} \times 60}{100} \Rightarrow 1{,}800 = \text{percent} \times 60 \Rightarrow \text{percent} = \dfrac{1{,}800}{60} =$

30, Therefore United States win 30% of medals.

9) Answer: D

A. $(3 \times 3.20) + (5 \times 0.90) = 9.60 + 4.50 = 14.10 > 11$

B. $(3 \times 2.24) + (6 \times 0.85) = 6.72 + 5.10 = 11.82 > 11$

C. $(5 \times 3.20) + (4 \times 2.24) = 16 + 8.96 = 24.96 > 11$

D. $(7 \times 0.85) + (3 \times 0.90) = 5.95 + 2.70 = 8.65 < 11$

10) Answer: C

We often describe the probability of something happening with words like impossible, unlikely, as likely as unlikely, equally likely, likely, and certain. The probability of an event occurring is represented by a ratio. A ratio is a number that is between 0 and 1 and can include 0 and 1. An event is impossible if it has a probability of 0. An event is certain if it has the probability of 1

impossible	unlikely	equally likely, equally unlikely	likely	Certain
0		$\dfrac{1}{2}$		1

11) Answer: D

The ratio of boys to girls in Maria Club: $16:64 = 1:4$

The ratio of boys to girls in Hudson Club: 1:4

1:4 same as 22: 88.

So, there are 88 girls in Hudson club.

12) Answer: A

Multiply the price by the sales tax to find out how much money the sales tax will add, then Add the original price and the sales tax.

State A: $45 \times 0.034 = 1.53; 45 + 1.53 = 46.53$

State B: $28 \times 0.045 = 1.26; 28 + 1.26 = 29.26$

Then take the difference: $46.53 - 29.26 = 17.27$

13) Answer: C

$\frac{3}{8} \times 16 = 6$

$\frac{1}{5} \times 60 = 12 \, min$

$\frac{6}{12} = \frac{x}{60} \rightarrow 12x = 6 \times 60 \rightarrow x = \frac{360}{12} = 30$ Ounces

14) Answer: B

Supplementary angles are two angles that have a sum of $180°$

$line \, R: 180° - 118° = 62°, Line \, S: 180° - 110° = 70°$

then in the triangle: $180° - (70° + 62°) = 48°$

$line \, P: \theta° = 180° - 48° = 132°$

15) Answer: C

average (mean) $= \frac{sum \, of \, terms}{number \, of \, terms} = \frac{(-25)+(-28)+(-21)+2+19+(-6)+38}{7} = \frac{-21}{7} = -3°F$

16) Answer: A

At least and Minimum – means greater than or equal to

At most, no more than, and Maximum – means less than or equal to

More than – means greater than

Less than – means less than Then, $P \leq 25$

17) Answer: C

$4 \times (-7) = -28$

18) Answer: B

A linear equation is a relationship between two variables, and application of linear equations can be found in distance problems.

$d = rt$ or distance equals rate (speed) times time.

$d = 1 \times 10 = 10$, then $(1,10), (2,20), (3,30)(4,40), \dots$

19) Answer: A

Let's find the mode, mean (average), and median of the number of minutes for each school.

Number of Minutes for school 1: 60, 60, 70, 70, 70, 80, 90, 90, 100, 110

$\text{Mean(average)} = \frac{sum\,of\,terms}{number\,of\,terms} = \frac{60+60+70+70+70+80+90+90+100+110}{10} = \frac{800}{10} = 80$

Median is the number in the middle. Since there are an even number of items in the resulting list, the median is the average of the two middle numbers.

Median of the data is $(70 + 80) \div 2 = 75$

Mode is the number which appears most often in a set of numbers. Therefore, there is no mode in the set of numbers. Mode is 70.

Number of Minutes for school 2: 50, 70, 80, 80, 90, 90, 90, 110, 120, 120

$\text{Mean} = \frac{50+70+80+80+90+90+90+110+120+120}{10} = \frac{900}{10} = 90$

Median: $(90 + 90) \div 2 = 90$

Mode: 90

20) Answer: D

$\frac{128}{4} = 32$

$\frac{256-128}{8-4} = \frac{128}{4} = 32$

PSSA Practice Test 2
Answers and Explanations

1) Answer: C

Divided 34 by 11: $\frac{34}{11} = 3.09090909 \ldots = 3.\overline{09}$

2) Answer: B

Day 3: $-0.70 \times \frac{2}{5} = -0.28$

Change price in days: $\left(75.85 + 0.52 + 0.66 + (-0.28) + (-0.70)\right) = 76.05$

Day 5: $76.92 - 76.05 = 0.87$

3) Answer: A

$1 hour = 60\ min \rightarrow \frac{1}{2} \times 60 = 30 min$

Rate: $\frac{\frac{5}{8}}{30} = \frac{x}{1} \rightarrow 30x = \frac{5}{8} \rightarrow x = \frac{5}{240} = \frac{1}{48}$

4) Answer: C

$2\frac{4}{15} \times \frac{-7}{15} = \frac{34}{15} \times \frac{-7}{15} = -\frac{238}{225} = -1\frac{13}{225}$

5) Answer: A

Hour: $x \rightarrow 51$ per hour: $51x$

Plus: add $(+)$, no more than: $\leq$; Then, $51x + 68 \leq 730$

6) Answer: B

By grid line: $d = 16$

Or distance for two points $(0,2), (0,-2)$: $d = \sqrt{(0-0)^2 + (8-(-8))^2} = 16$

$d = 16 \rightarrow$ Circumference $= \pi d = \pi(16) = 16\pi = 50.24$

7) Answer: D

3 hours equal 6 half hours

$6 \times 3° = 18° \rightarrow 8 - 18 = -10°F$

8) Answer: C

If the price of a printer is decreased by 43% then: $100\% - 43\% = 57\%$

$57\% \; of \; h = 0.57 \times h = 0.57h$

9) Answer: C

The reflection of the point (x, y) across the y-axis is the point $(-x, y)$.

If you reflect a point across the $y-$axis, the $y-$coordinate is the same, but

the $x-$coordinate is changed into its opposite. Reflection of $(1,5) \rightarrow (-1,5)$

10) Answer: B

$-5x + 12 > -8$ (Subtract 12 both sides)

$-5x > -20$ (Divide both side by -5, remember negative change the sign): $x < 4$

11) Answer: D

$25v + 2,340 = 11,840$ (subtract 2,340 from both sides)

$\rightarrow 25v = 9,500$ (divide both sides by 25) $\rightarrow v = \$380$ cost of each table.

12) Answer: B

Difference of temperature is: $|t_2 - t_1| = |16 - (-7)| = |16 + 7| = |-7 - 16|$

13) Answer: A

Supplementary angles are two angles with a sum of 180 degrees.

$\alpha + \beta = 180°$ and $\beta = 125° \Rightarrow \alpha = 180° - 125° = 55°$

complementary angles are two angles with a sum of 90 degrees.

$\alpha + \gamma = 90$ and $\alpha = 55° \Rightarrow \gamma = 90° - 55° = 35°$

14) Answer: D

A. Expenses in Year 2: $2,000 \rightarrow 2,000 + 500 = 2,500 \neq$ Year 3

B. Income in Year 1: 1,900 and $20\% \; of \; 1,900 = 380 \rightarrow 1,900 + 380 =$

   $2,280 \neq$income of Year 5

C. Income's year 3, 4, and 5: $1,800 + 2,000 + 2,500 = 6,300$

   Expense's year 2, 3, and 4: $2,000 + 1,500 + 1,700 = 5,200 \neq 6,300$

D. Half of Year incomes 4: $\frac{2,000}{2} = 1,000 < 1,500$ Expenses in Year 3

15) Answer: C

$$(4n - 7) - \frac{1}{2}(11 - 8n) + \frac{7}{2} = 4n - 7 - \frac{11}{2} + 4n + \frac{7}{2} = 8n - 9$$

16) Answer: C

If you ever need to find the percentage of something you just times it by the fraction. So, all you need to do to figure this out is to find 45% of 12.14 which is approximately 5.46.

17) Answer: C

$\angle D = \angle E = b° \rightarrow FD = FE = 3cm$

Perimeter of triangle: $3 + 3 + 5 = 11\ cm$

Actual triangle: $11 \times 5\frac{1}{5} = 11 \times \frac{26}{5} = \frac{286}{5} = 57\frac{1}{5}\ cm$

18) Answer: B

The ratio of boy to girls is 3:4. Therefore, there are 4 girls out of 7 students. To find the answer, first divide the number of girls by 4, then multiply the result by 7.

$24 \div 4 = 6 \Rightarrow 6 \times 7 = 42$

19) Answer: D

1 L=1,000 mL

1 fl oz = 29.57 mL$\rightarrow$ 1 fl oz =0.02957 L

$88.71\ L = \frac{88.71}{0.02957} \times \frac{100,000}{100,000} = \frac{8,871,000}{2,957} = 3,000\ fl\ oz$

20) Answer: B

$$4(x + 6) + (x + 10) + 46 = 180 \rightarrow 4x + 24 + x + 10 + 46 = 180$$

$$5x + 80 = 180 \rightarrow 5x = 100 \rightarrow x = \frac{100}{5} = 20 \Rightarrow x = 20°$$

$$\angle B = 4(x + 6°) = 4(20° + 6°) = 104°$$

"End"